THE CULTIVATED LANDSCAPE

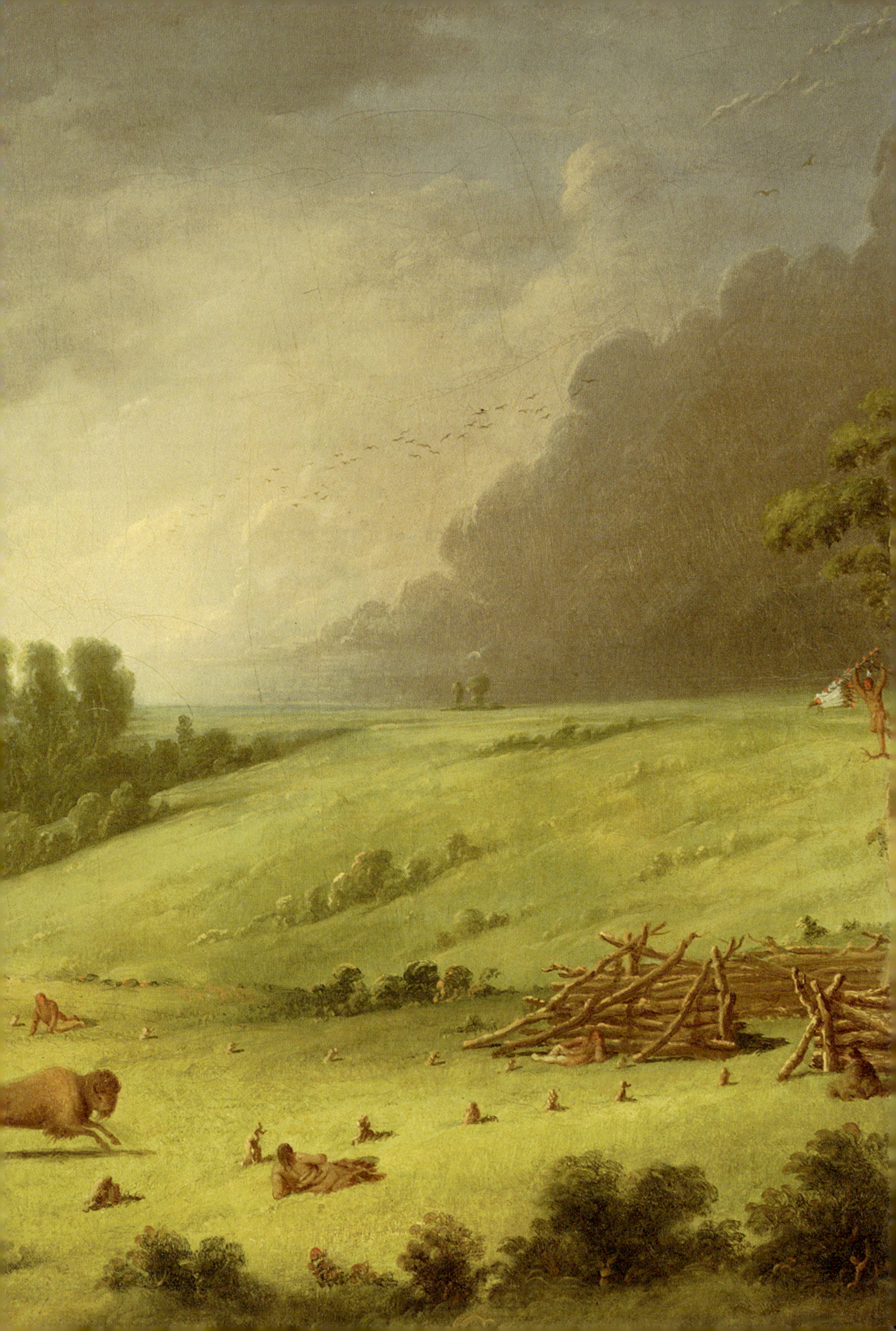

THE CULTIVATED LANDSCAPE

An Exploration of Art and Agriculture

CRAIG PEARSON & JUDITH NASBY

McGILL-QUEEN'S UNIVERSITY PRESS | Montreal & Kingston · London · Ithaca

ISBN 978-0-7735-3246-5

Legal deposit third quarter 2008
Bibliothèque nationale du Québec

McGill-Queen's University Press acknowledges the support of the Canada Council for the Arts for our publishing program. We also acknowledge the financial support of the Government of Canada through the Book Publishing Industry Development Program (BPIDP) for our publishing activities.

Library and Archives Canada Cataloguing in Publication

Pearson, C.J.
The cultivated landscape : an exploration of art and agriculture /
Craig Pearson & Judith Nasby.

Includes bibliographical references and index.
ISBN 978-0-7735-3246-5

1. Agriculture in art—History. 2. Agriculture and the arts—History.
I. Nasby, Judith, 1945– II. Title.

ND1460.A39P42 2008 704.9'4309 C2007-905192-8

This book was designed and typeset by studio oneonone in Sabon 10.5/14

Frontispiece: Paul Kane, *A Buffalo Pound*, 1846 (oil on canvas)
Collection: Royal Ontario Museum (ROM 912.1.33) www.rom.on.ca
Photo: With permission of the Royal Ontario Museum © ROM

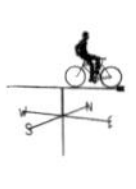

Image used to indicate section breaks:
Rodney Graham, *Weather Vane*, 2002 (black enamelled stainless steel).
Collection: Purchased with funds donated by Helen Brimmell, 2002,
Macdonald Stewart Art Centre Collection, www.msac.ca

CONTENTS

PREFACE

Agriculture is a significant part of human civilization not only because a primary measure of its success is the provision of sufficient and safe food, but also because its use of land has created, both consciously and accidentally, the landscapes we know. Artists have described and commented on these created landscapes, and this book is illustrated with artworks that reveal the fascinating changes that have taken place on the historical journey to the present. As the paradigms of agricultural practice have shifted, so has the relationship between art and agriculture. The idyllic depiction of manorial landscapes in the eighteenth century have, in our own age, given way to artistic expressions of dislocation.

The issues facing agriculture today are complex. The authors' objective is to bring these issues to the attention of the public and of academic communities and to encourage discussion of possibilities for the future. The book examines six eras of Western agriculture as it evolved from the Middle Ages to the present; its geographical and historical context is therefore vast. As Lytton Strachey wrote in the preface to one of his own historical explorations, the authors have "rowed out over the great ocean of material and lower[ed] down into it, here and there, a little bucket, which will bring up to the light of day some characteristic specimen, from those far depths, to be examined with a careful curiosity."

ACKNOWLEDGMENTS

We would like to especially thank Dawn Owen, assistant curator, Macdonald Stewart Art Centre; Delia Bucknell, Ontario Agricultural College; and Anne Marie Todkill, who provided invaluable support in research and editing.

Craig Pearson sincerely thanks the Rockefeller Foundation, New York, for providing a fellowship that allowed him time to write at Bellagio. Thanks also for access to the Rockefeller Archives and to Thomas Rosenbaum, archivist, The Rockefeller University.

Judith Nasby acknowledges Macdonald Stewart Art Centre staff members Sorouja Moll, Verne Harrison, and Aidan Ware for their ongoing commitment. Thanks also to the Ontario Arts Council and the Canada Council for the Arts for supporting scholarship by contributing to this publication.

THE CULTIVATED LANDSCAPE

1

CUSTODIANS OF THE LANDSCAPE IN EUROPE

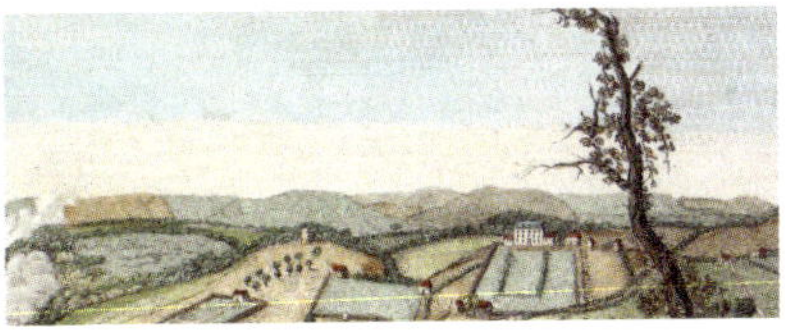

On Lake Como, in northern Italy, summer days start clear: the water is blue, the houses crisp orange and white against the grass and pines on the steep hillsides. By mid-morning the haze rolls in, obscuring the crispness. At Cernobbio, Villa d'Este stands beside the lake, its rendered walls ornamented in ochre, orange, grey, and cream hues. This sixteenth-century villa (1568) is a fitting scene for the beginning of our story: landscapes and land-use in Europe before the Agricultural Revolution. Formal lines of cypresses cut across handmade terraces that would once have supported vines, herbs, and vegetables. The main lines of the Renaissance gardens still exist, including two huge olive trees and a 500-year-old plane tree – the botanical landmark of the villa at about 9 metres in circumference and 40 metres high. When Caroline of Brunswick, Princess of Wales, purchased the villa in 1815, she introduced picturesque elements that forever changed the garden's aesthetic (Fig. 1). Stone walls screen out the surrounding fields; once allotted to peasant's crops and livestock, they now hold modest houses and waste land. An avenue of cypresses and waterfalls and the nearby fields are reminders of the manorial era of land custodianship in Europe, to which we will return.

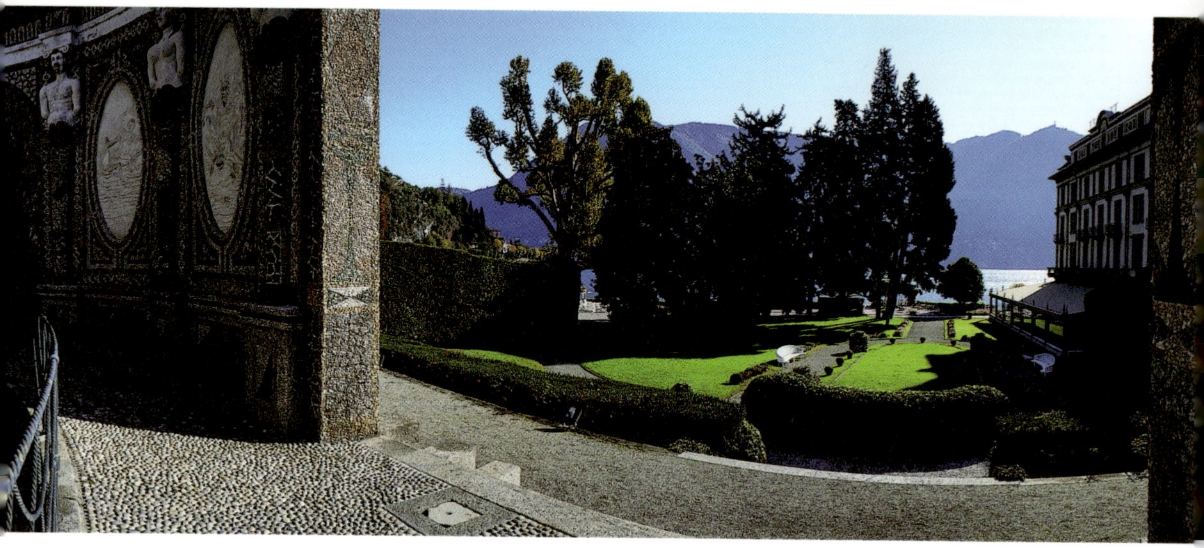

Figure 1
The Nymphaeum at Villa d'Este, Cernobbio, Italy, exemplifies the blend of classical and picturesque aesthetics in the gardens at Villa d'Este. Elaborate parterres and a perspective avenue ascend to the Nymphaeum, also called "the Mosaic" in view of its decoration with polychrome pebbles. Picturesque trees border its manicured grounds, an example of the elements added after 1815 by Caroline of Brunswick.

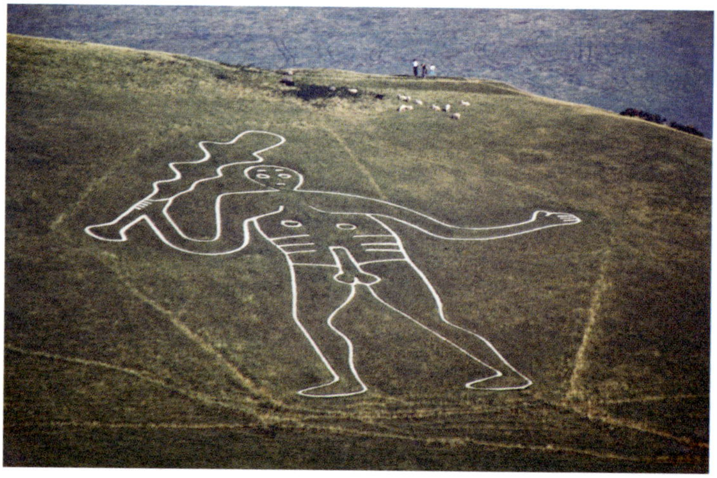

Figure 2
Cerne Abbas Giant, Dorset, England, circa AD 180–193 (disputed). Fifty-five metres long and formed by a 0.3-metre trench cut into the chalky hillside near the Dorset village of Cerne Abbas, the giant has been considered a representation of the Roman god Hercules, who over time became a fertility symbol. May Day celebrations associated with fertility rites have been held on the site for centuries.[1]

Once upon a time in Europe there were isolated settlements based on shifting "slash and burn" cultivation, and most of the landscape was covered with native vegetation. (Today, this type of land use persists only in isolated pockets of the world; it was fairly common until the 1980s in Sumatra and other less populated islands of Indonesia, in Latin America, and in parts of Africa.) The people who subsisted on this shifting cultivation were usually governed – or dominated – by warlords. They were also governed by religious and superstitious practices that gave meaning to seasonal planting and harvesting cycles and were believed to help ensure the fertility of the land and the people. An example is the *Cerne Abbas Giant* (circa AD 180–193), a survivor of early spirituality in Britain (Fig. 2).

With the passage of time, Europe's early warlords evolved into a land-owning class, feudal lords who oversaw an agriculturally based manorial system. Their houses, the manors, had walled gardens that produced vegetables and flowers and whose designs derived from Greek and Roman times (Fig. 3). Large monasteries and palaces such as Charlemagne's at Aachen (AD 792–805) dominated the cultivated landscape. In 1533, the household garden moved outside the walls, coining the term "outdoor" with the design of a garden overlooking the Thames at Hampton Court.

However, it was beyond the walled gardens of the manors, palaces, and monasteries where human activity had the greatest impact on the landscape. Manors and clusters of hut dwellings for peasants and animals were embedded in widely cultivated landscapes. The lords possessed both land and serfs. The possession of serfs entailed some responsibilities for the lord, as described in the *Sachsenspiegel*, an oft-quoted thirteenth-century document from southern Germany: "We shall serve our lords so that they must protect us. If they do not protect us we are not bound by the law to serve them."[2]

However, the power was held by the lords, who determined the peasants' obligations to them as well as their entitlement to the use of land for agriculture and their access to woodlands and grasslands. On a typical feudal estate in the Middle Ages, most farming was done by serfs, who were obliged to provide labour: these, the *servi* or *mancipia*, had the lowest social status on the estate. As Rosener recounts from an early tenth-century land register from the

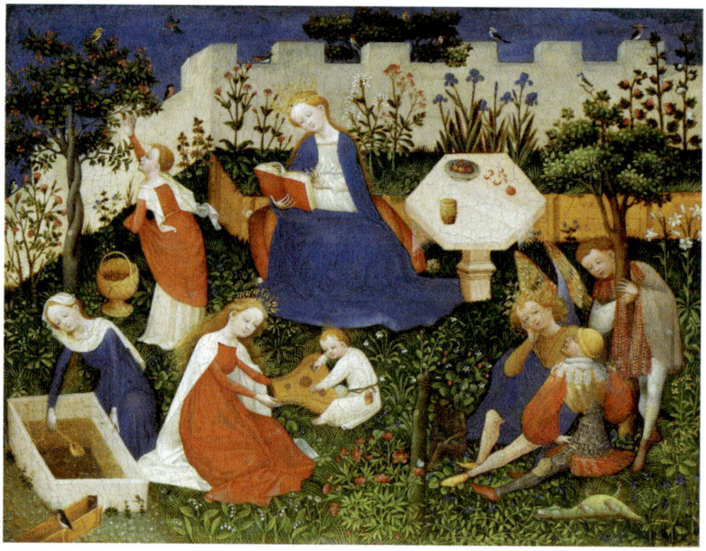

Figure 3
Master of the Middle Rhineland, *Virgin in a Garden,* circa 1410 (tempera on wood). A gothic walled garden is shown in this 15th-century painting, a fantasy that depicts the Virgin Mary surrounded by saints lounging like courtiers. St. George is shown with his docile dragon belly-up, basking in the sun. The tranquility of this private paradise is emphasized by the manicured flowers and cherry trees.

imperial monastery on the lower Rhine, there were 121 scattered holdings farmed by dependent peasants, who owed services to the lord. Peasants were divided into three main groups (Fig. 4). Serfs, who had no holdings of their own, accounted for about 25 per cent of the lord's dependents and were obliged to perform services at his will. Tenants of holdings were also obliged to provide services, while the third group, peasants, paid rent without an obligation of labour. Tenants, by far the largest group,

were obliged to perform sowing work for two weeks of sowing each spring and autumn, which meant that they had to plough about 3.6 acres and prepare them for actual sowing of the seed. An additional two weeks of service was due in June, and they were obliged to help with the hay and grain harvests, to undertake various tasks in the lord's garden and to aid in the mustering of the manorial pigs. All in all, the service of these tenants amounted to ten weeks, hence a fifth of the year. In addition to this considerable workload they had to pay fifty-six pfennigs (Austrian silver coins minted in the twelfth century), 3 chickens, 10 eggs and 12 bushels of grain.[3]

The third group of peasants enjoyed a somewhat higher status. They paid their dues to the lord either as money or in kind, and their burden was smaller. Peasants must have struggled to meet the demands of the lord: it is clear that it was their labour, not the land,

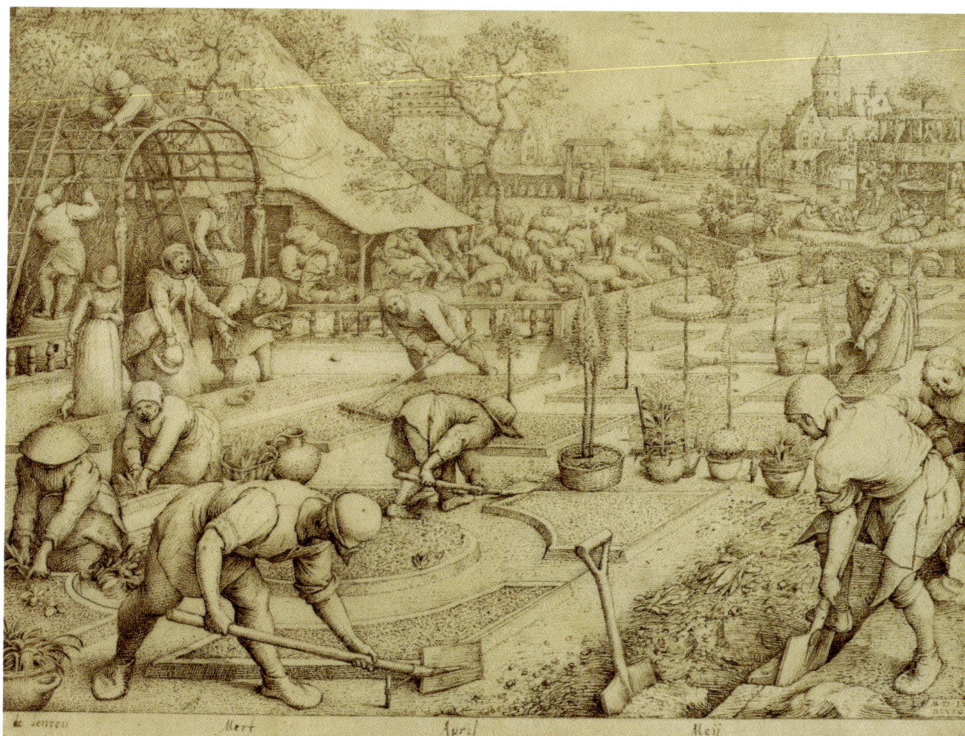

Figure 4
Pieter Brueghel the Elder, *Der Frühling (Spring)*, 1565 (brown ink on paper). In this compressed view of manorial life, we can observe many different activities. The lady shown on the left beside the man sowing has left her castle to inspect the planting being done by her serfs, who are creating formal raised parterres for vegetables and flowers. The serfs are potting shrubs, ready for transplanting to create formal geometric designs. In the background shepherds are shearing sheep and tending calves. Meanwhile the lords and ladies (upper right) cavort in their private garden below the castle, enjoying the fruits of the serfs' labour.

that limited the productivity of the system. Indeed, the system persisted, for the benefit of the lord, because the early Middle Ages in Europe had been a time of turmoil. Lawlessness and robber-lords abounded, and towns were depopulated. The population of Rome, for instance, declined from one million people in AD 300 to fewer than fifty thousand in the Middle Ages. The lords controlled the amount of land in agricultural use. A basic measure of land was the hufe, hide, bol, or mansus, which described a normal peasant holding. The size of a holding depended on the fertility of the area, as it was intended to be adequate to support a peasant and his extended family; usually it was about five hectares, but could be as much as thirty.

The land that supported each peasant family was embedded as strips within collectively managed fields scattered about the manor. These strips were arranged in a regular fashion within great, open fields, often ordered in the same sequence as the houses (or hovels) along the village street (Fig. 5). A neighbour in the fields would be next door in the village, too.

Most cereals were grown in a strictly ordered "three-field" system. Two crops were grown in sequence: commonly rye (throughout much of central and northern Europe) or wheat (in milder areas) or spelt, followed the next year by oats or barley. Then the land was fallowed (either ploughed or left idle) until the rotation started again in the fourth year. As with the crops, the how and when of ploughing, sowing, and harvesting were prescribed by the lord or his overseer.

These prescriptions were carried from generation to generation in tapestries, writing, or oral devices such as poetry. Columella (Lucius Moderatus of Cadiz, who lived to about AD 70) records early instructions that would have remained unchanged into medieval times, such as, "Beet is put into the ground in the form of seed when the pomegranate flowers, and as soon as it consists of five leaves it is, like cabbage, transplanted," and "It is correct to sow the radish twice a year, in the month of February, when we expect the spring crop, and in the month of August about the time of the festival of Vulcan."[4]

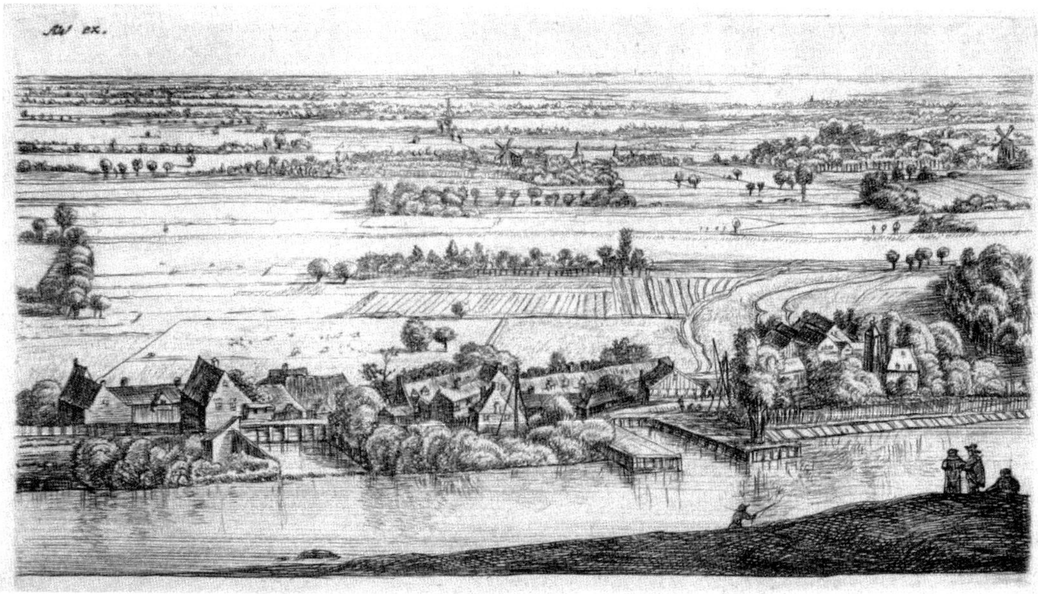

Figure 5
Antoni Waterloo after Johannes Ruisher, *Village on the River,* circa 1640 (drypoint and etching). This etching gives a panoramic view of the Rhine River from the tower of the church of St. Cunera in the village of Rhenen. Strip fields are located behind the houses that line the edge of the river. Sluices in the dykes channel water to the fields that are used for growing grain.

Depictions of peasant occupations were widespread in the form of sculptural works on the exterior of Romanesque cathedrals, and in illuminated manuscripts. Illustrated psalters were made for the daily devotional requirements of the owner. They also contained monthly calendar illustrations of seasonal agricultural activities. An example is *October* in the Fécamp Psalter, a Flemish manuscript dating from the twelfth century (Fig. 6). Medieval peasant illustrations often alluded to scripture, as in the story of Adam and Eve. Adam became a cultivator of the soil and Eve a weaver of cloth after their expulsion from the Garden of Eden. For illiterate peasants these images, and their Biblical exemplars, guided the cyclical rhythm of their seasonal work.[5] Folklore, remembered through devices such as rhyming couplets, reminded lord and tenant alike as to the timing of agricultural operations and the dependence of agriculture on the seasons: "Clouds throughout the day, A clear sky at night: Famine." Easily remembered maxims also dispensed advice to the next generation, whether in western Europe or northern India: "A spendthrift son, A cross-eyed buffalo, A moody ox: Get rid of them at once."[6]

For thousands of years the implements of farming did not change: plough-lines and mounds at the edges of fields accumulated from Neolithic times through the manorial era along elongated strips of land, shaped through the centuries by the single-furrow plough, which moved the earth ten to twenty centimetres with each pass. An example of such a plough – long-shafted, wooden, and believed to date to 2000 BC – is housed in an archaeological museum in Desenzano, near Verona; such ploughs would have been pulled by oxen, horses, or men from the beginning of cultivation, right through to the widespread adoption of tractors in the 1950s.

In Europe, by about the twelfth century, regional specialization of grain production had evolved, rye and spelt being grown in colder areas. Specialization fostered trade between regions. In some specific areas, because of good soil or abundance of manure, there are records of single, continuous cropping: the farmers of Eschboden, in northwest Germany, grew rye continuously for centuries, maintaining the fertility of their fields by bringing in topsoil, sod, and animal (and no doubt human) manure.

During the manorial period livestock were small and slow; cattle, oxen, and horses stood half- or two-thirds as tall as we are used to, in the twenty-first century. Humans were shorter, too, in those times (Fig. 7). The lord usually retained control of the breeding and housing of bulls, and sometimes of stallions and boars. The peasants, however, had the task of looking after most animals. Pigs and chickens were favoured because they quickly grew to a stage when they could be eaten, and because they would eat almost anything. Before slaughter, pigs would be fattened in woodlands, where they fed on acorns and other wild material. (Pig-rearing in the medieval manner persists regionally today in Corsica, where boars are kept in transient pens made from forest timber, at roadside bends or other spots that allow their easy release into the forest for fattening.) Sheep were also popular in the manorial system because they pro-

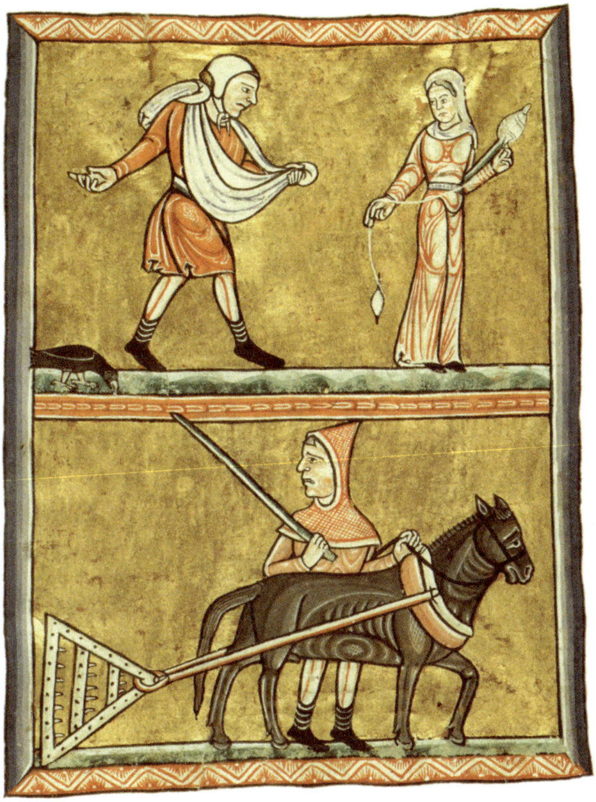

Figure 6
October, from the Fécamp Psalter, circa 1180 (illuminated manuscript). In the upper register a man sows and a woman spins with a distaff and spindle. In the lower zone a man uses a horse-drawn harrow with spikes to break the soil.

vided wool and meat as well as milk; being of smaller size and value, they were also more affordable than cattle, for example.

Historically, oxen were the main form of draft animal. However, horses became increasingly popular; they were smaller and cheaper than oxen, especially in England where, there being little demand for old horse-meat, they devalued with age more than oxen did. Horses were also faster, if the terrain was reasonable. Hence, all-oxen ploughs in early tapestries give way to horses leading, perhaps, four oxen, and then to all-horse ploughing, which might carry out the job at twice the speed. Horses were especially favoured by peasants, when they could afford any draft animal at all, because they provided quick transport as well as draft power.

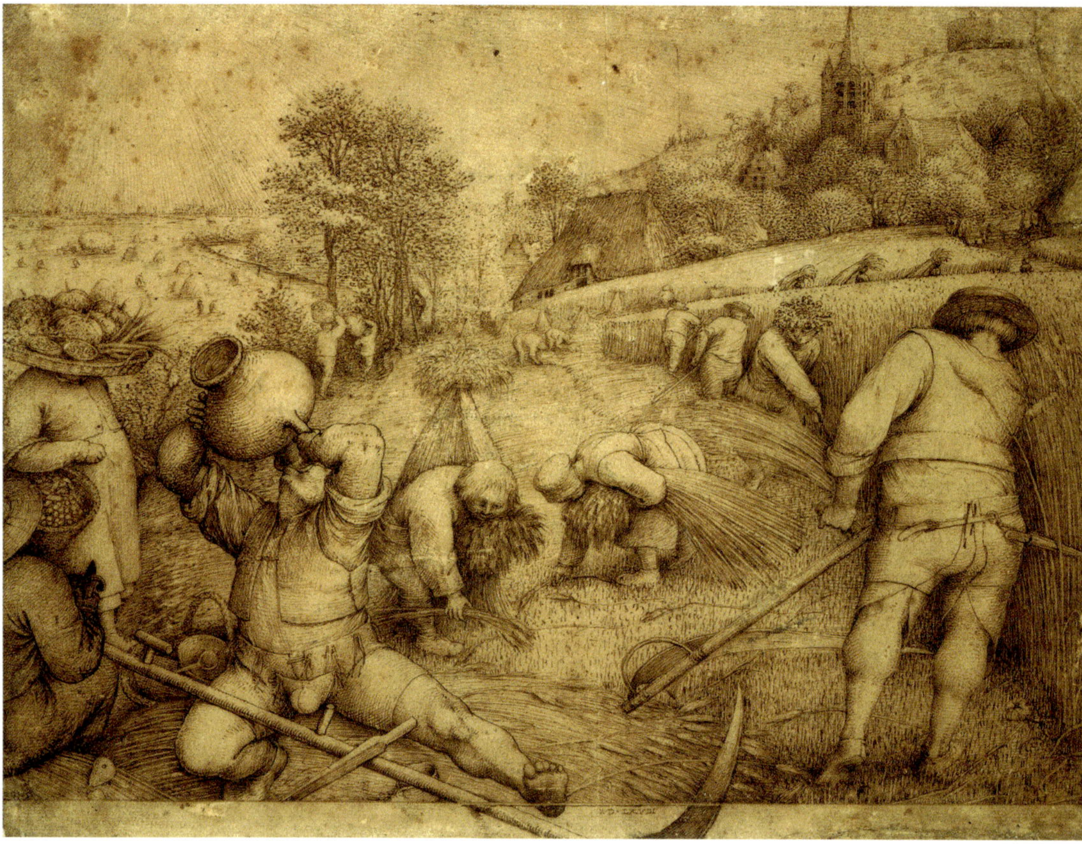

Figure 7
Pieter Brueghel the Elder, *Summer – Peasant Harvest,* 1568 (pen and brown ink on paper). Brueghel celebrates the success of the August harvest of rye in the Lowlands with a caricature of peasant life and a depiction of a bountiful harvest – a demonstration of success and contentment on the manor. A powerful reaper scythes grain beside a comical depiction of a man drinking from a jug. The women gather wheat sheaves to form stocks, and in the far background a team of horses pulls a hay wagon. A female figure appears on the right, supporting a basket of vegetables on her head and carrying a basket of produce.

Through the eleventh to fourteenth centuries the system in Europe was changing. The rapid and sustained rise in population was so great that some writers have calculated that, had it continued, Europe would have attained a population of one billion by the early part of the twentieth century. Rising population shifted the balance of scarcity. Lords increasingly preferred money, rather than labour, as rent. The rise in population also increased settlement and encouraged the clearing of native vegetation for permanent dwellings and permanent fields, rather than rotation (or slash-and-burn and then abandon) agriculture. The increase in available labour, and the growth of old towns and new, gave rise to regional migration and began the process of urbanization that continues today throughout most of the world.

Two hundred years of population growth and large-scale clearing of woodlands caused agriculture in Europe by about 1300 to extend to higher-altitude and rougher terrain than before or since: the extent of farming has contracted ever after. By this time deforestation must have had widespread direct and indirect effects, in addition to changing the appearance of the landscape. Peasants relied on trees for firewood, charcoal, and for fruits, berries, and nuts to eat. A shortage of grazing lands caused lords to restrict the use of the commons and of alpine grasslands – and, sometimes, the number of livestock that peasants were allowed to keep. The loss of trees would also have irreversibly changed the hydrology of the region, causing more runoff of rainfall and a loss of soil nutrients. Various authors have suggested that these effects, together with the steep increase in population, contributed to the famines and local desertification that occurred throughout Europe, beginning early in the fourteenth century. Although some of these areas have been reforested, especially those that are isolated or at high altitudes, others remain as scars inflicted by the march of civilization: on a summer's day, travelling by train from Rome to Naples, one might see the sun bouncing off eroded yellow-red hill slopes that, first cleared in the Middle Ages, remain unproductive, eroded of topsoil, to this day.

Population growth was checked intermittently not only by overuse of the land in local areas but also by a series of poor harvests and famines in 1315–17 and then by the Black Death in 1348. These, and less spectacular failures, resulted in large, intermittent migration as well as falling birth rates. Further, two major changes in land management affected the style of agriculture and the landscape of Europe. One was the dissolution of the monasteries, which occurred at various times since about 1100 throughout continental Europe, and in England and Wales in 1536–39, and which threw one-fifth of the land, and the labour supply associated with it, onto the market. The second was the aggregation, or enclosure, of fields.

An increasing population caused peasant holdings to be split into smaller and smaller strips as they were passed from father to son. This in turn was a stimulus toward better management and the intensification of grain production. It also shifted the supply of labour such that it was no longer precious: plentiful labour meant cheap labour and encouraged lords to extract payment, rather than labour, from peasants. This, in turn, allowed lords to hire employees, thereby being able to select some specialization of labouring skills, and specialization of the crops grown. All these trends shifted land use away from collections of strips scattered about open fields and regulated by a lord, toward the cultivation of large, enclosed fields.

The term "enclosure" can be used to refer to how landowners, whether lords, wealthy merchants, or farmers, enclosed or marked out their fields as contiguous holdings in a permanent fashion, establishing their boundaries with low stone walls, fences, or hedgerows (Fig. 8). The term "enclosure" is also used to describe the change of ownership and in management that took place, often under the force of law, from scattered plots in large, collective fields, to consolidated fields and top-down management. Consolidation has occurred throughout history, but the intensive period of enclosure dates from about 1600 in England and continued until the late 1780s in the lowlands of Scotland and to the 1820s and 1830s in the Scottish highlands and islands. In continental Europe, the enclosure process was more gradual and, in general, occurred somewhat later. Thin strips of fields are still common in parts of Germany, Austria, and Poland today, and the landscape of the Czech Republic is in some places characterized by a patchwork of strips and, in others, by large fields created by Communist collectivization and post-1993 aggregation in a market economy.

Enclosure by agreement was the most common, and the earliest, method of agricultural consolidation. Genuine agreement or coercion of the peasants resulted in consolidated holdings, particularly in south and central England, and in some cases special arbitrators were appointed to oversee the process for the benefit of all. For example, one petition by the owners and customary tenants of a Northumberland township called for the appointment of commissioners and referred to "the daylie inconvenience which they suffer

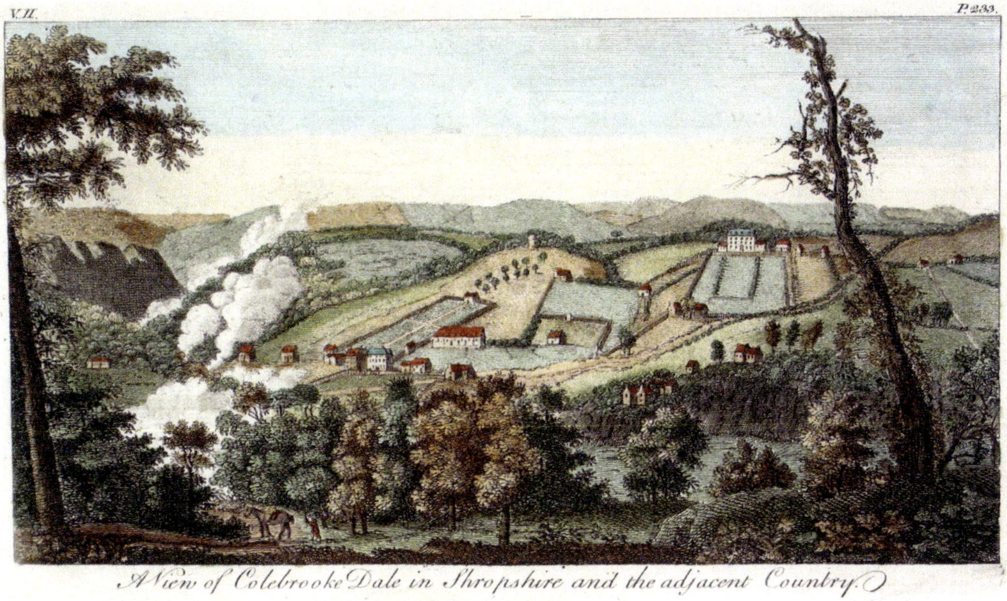

Figure 8
Artist unknown, *A View of Colebrooke Dale in Shropshire and the Adjacent Country*, 1776 (hand-coloured copper engraving). This illustration from the publication *A New Display of the Beauties of England* depicts fields stoutly enclosed by stone walls and hedgerows. The image announces the prosperity of the hamlet, a key site for the start of the Industrial Revolution, which is signified in the engraving by smoke rising from industries located in the valley. The introduction of mechanized steam-driven machinery had a dramatic affect on the economic and social structure of agrarian societies.

by reason of the intermixture of their lands and tenements." Another justified the call for change on the grounds that under the existing system of scattered and shifting fields farmers "cannot make any improvement."[7]

Voluntary enclosures were piecemeal but widespread: even if the fields were not marked out with hedges or in some other manner, the voluntary relinquishing of common rights of use had caused much of the land management to be consolidated in counties such as Wiltshire, Dorset, and Somerset. Writers held up this aggregation of land as a reason for the prosperity of these areas. Half or more of England's farmland may have been enclosed by 1675 or 1700. After this, parliamentary acts forced consolidation on the peasantry. The most ruthless consolidations occurred late in this period, when rents were raised dramatically in the Scottish highlands in the 1770s to "voluntarily" displace small farmers so that landlords could run more sheep. Small holdings (of about five acres) were no longer sufficient to maintain a farm family. This tactic was followed by eviction, and consolidation. In the winter of 1828–29, 1500 people, or 20 per cent of the farmers on the Isle of Arran, were forced off their land.[8] A small number – fifteen or twenty – families migrated to Canada and contributed to the agricultural expansion there, but, as in other such cases of displacement, most drifted to other areas in their own country as agricultural labourers or unskilled town workers.

Whether it occurred by mutual consent or by government-backed force, enclosure had the effect of creating larger agricultural units. The average farm increased to 145 acres by about 1800, indicating that, in England at least, farmland "had effectively passed from family farmers to large-scale capitalist tenants."[9] Because these farming units were large enough, and did not depend on coerced labour, they and their owners or overseers could benefit from innovation. Tenants, who were now paying rent rather than tithed labour, also had a greater incentive and perhaps more time to be innovative.

Both the dissolution of the monasteries and the process of enclosure produced a large landless class of agricultural labourers and the urban poor. During the eighteenth century, Europe's population increased from 102 to 154 million. Greater population growth (0.8% per year compared with 0.5% for continental Europe), earlier enclosures, and customs of inheritance in England created a landless class more quickly there than in other European countries. It also caused more rapid migration to towns; England's urban population rose from 13–16 per cent in 1700 to 22–44 per cent in 1800. By contrast, in continental Europe the percentage of the population living in towns was only 12 per cent in 1800. Hopcroft[10] describes, as well as can be done in retrospect, the farming systems and tenancies in Britain and continental Europe until about this time. Regional variations within each country were substantial, but inter-country differences were great too: in France, where land was divided when it was passed from one generation to the next, plots became smaller and smaller, although this did not stimulate much migration to towns, unlike in Britain, Germany, and the Netherlands, where the right of primogeniture (inheritance by the oldest son) kept farms intact and displaced those who did not inherit.

Although the landscape had become more dynamic, and holdings were of a larger scale as a result of the enclosure movement, the cycle of life nonetheless had changed little in thousands of years. The cycle of collective hand sowings and harvesting by peasants continued (Figs. 6, 7). Similarly, a closer look at the manor-houses in contemporary engravings and tapestries would show that they have not changed much since the era began. Walk inside the engravings: walk into a chateau or estate-house of the 1400s, which you might find in present-day France. The walled gardens still contain fruit trees and vegetables and flowers, while outside the walls hired labourers work in fields of wheat or rye.

Artists have reinforced the notion of a changeless agriculture with recurring images of men ploughing fields using single-furrowed ploughs pulled by oxen and horses, women in groups sowing seed by hand, and women and men gathering in the harvest. The similarity of detail between Vincent van Gogh's (1853–1890) depictions of harvesting and its representation in Figure 5, from two centuries earlier, is probably accurate, if perhaps a bit wistful. Nonetheless, this depicted changelessness concealed technical innovations in agriculture that were to affect land use and, ultimately, the landscape. Most of these innovations originated in England and continental Europe, and many were trialled, if not widely used, in British agriculture before 1750. For example, the Rotterdam plough was introduced in the 1760s, and mechanical seeding devices ("drills") were first patented in 1623 and broadly trialled in the 1680s and 1690s (although Jethro Tull has traditionally been credited with the invention of the seed drill in the 1730s). Devices to thresh the grain from the rest of the head after harvest were first patented in the 1732. (Mechanical harvesters were not trialled until 1826, and satisfactory machines, developed by Americans Cyrus McCormick and Obed Hussey, were not displayed in Europe until the Great Exhibition of 1851.)

It is a matter of academic conjecture as to how widespread the impact of these innovations actually was. Although the art of this period might give a somewhat skewed view through a preference for more interesting hand-farming scenes over the representation of less aesthetically pleasing machines, it is unlikely that the use of these devices was widespread beyond the larger, more innovative manors until the nineteenth century. Indeed, although New World agriculture became mechanized rapidly, much of the smallholder and tenant-farmer agriculture in Europe and elsewhere in the world, such as in Asia, relied on manual labour rather than technology into the mid-twentieth century.

Despite the more widespread reality – that is, a relatively open landscape, such as that depicted in Figure 4 – many paintings from the seventeenth and eighteenth centuries captured a serene, largely wooded countryside. From Claude Lorrain's view of Campagna near Rome (circa 1650), to Johann Brand's *The Danube Near Vienna* (1790) and John Constable's *Hampstead Heath* (1821), nature is wooded and arcadian. Thomas Gainsborough, early in his career (from 1747 to 1759), produced landscapes in the style of the French pastoral idyll; later, moving to London, he found portrait-painting to be more profitable. Thus, the art movement in the Western world shifted from renaissance and baroque conventions to neoclassicism and romanticism.

"Naturalness" and romanticism in art reflected, and reinforced, an era of creativity in landscape design. But there were at least two other influences. One was a revival of the classic Greek and Roman villa landscapes. The first of these was in Siena in 1338–40, and this architectural trend became popular in Italy and France in the sixteenth and seventeenth centuries. Inspired by classical influences, landscapes were designed as an extension of the house's architecture into its surroundings. In the Renaissance, Italy showed the way in the art of the gardens, with highly ordered plantings of hedged shrubs, flowers, herbs, and vines, all arranged in carefully designed, and counted, spots. These were framed with cypresses, pines, and elegant waterfalls and fountains. Geometry dominated, usually with plantings radiating outward from the villa or arranged in a grid. The Villa d'Este near Como exemplifies the renaissance garden on the scale of the large country home, the reward of the lesser noble, cardinal, or businessman (Fig. 1). The villa was built by Pellegrino Tibaldi for a cardinal in 1570. Between the cardinal's grand house and the more recently built princess's villa to the right is the centrepiece and entry to the gardens: a nymphaeum faced with coloured stone-chip mosaics, through which one enters an avenue of tall dark cypresses that look as old as time. Between the cypresses are two gentle waterfalls, dropping through a series of stone and cement troughs; the formal avenue ends, up the hill, with a statue of Hercules fit for pre-Christian Rome. On the right, the main road cuts beneath the cypress avenue and across what would have been gardens and paths leading to fake fortifications around which, from 1810, guests of Napoleon's aide-de-camp, the then owner, fought mock battles. On the left of the avenue, the hillside was once terraced and planted according to a meticulous design. Today a metre-wide grindstone, for crushing olives, stands irrelevantly near tennis courts created by chopping apart the terraces and formal plantings. Ironically, what remains of the Villa d'Este is the result of its transformation into a luxury hotel.

Formality in the French or Italian revival style is still accessible, however, and can be seen on a grand scale in the public gardens that were laid out in the country or summer houses of the monarchs of Europe at Versailles near Paris, Schönbrunn and Schloss Niederweiden outside Vienna, Drottningholm outside Stockholm, and even at the Troja Palace in the countryside outside Prague, where the hapless Archduke Ferdinand spent most of his time among the roses, far from the *gravitas* of the imperial throne in Vienna. Renaissance gardens were carefully planned: each dot on the plan might represent marjoram or thyme, each individually drawn tree an olive, pear, or lemon. All elements were counted.

Both formality and exuberance were evident at Versailles. The original water basin was excavated in 1639; drawings of the gardens date from 1652–61 and were elaborated until about 1693. The scale was truly monumental. The construction of Versailles involved the digging of a canal 5200 by 400 feet, the razing of three villages, and the engineering of awe-inspiring fountains. "By the end of the first phase of the development of Versailles in the summer of 1668 when the so-called Great Fete was presented in August, full-scale water displays were regular features of garden entertainment, consuming more water in a day than the pumps of Samaritaine delivered to the entire population of Paris of 600,000 people."[11] André Le Notre's gardens at Versailles combined formal interlocking designs, providing highly patterned and separate spaces for people and plants. Even today, after the gardens have been stripped of some of their exuberance and simplified, Versailles has 200,000 trees and employs 600 people to maintain the grounds and the palace.

A slightly more recent example, in Britain, is the garden at Badminton (Fig. 9), an extensive complex of geometric designs dating from the early eighteenth century. A second and later influence was exquisite, formal minimalism from Asia, particularly China and Japan. Buddhism was brought from China via Korea to Japan in the sixth century, and in AD 794 the Japanese capital was moved from Nara to Kyoto, known as the "Capital of Peace and Tranquility." It is not known whether the first Zen garden dates from this time; it is recorded that Rankei Doryu created a Zen garden for the temple Kenchoji, now called Tokoji, in 1253. The greatest of the Zen master garden designers was Muso Soseki (1275–1351). Significantly, elements of the design were first painted in ink on paper, than transposed into gardens. Landscape painting in China and Japan had an important influence on Japanese garden design.

The formal, walled house-garden or *giardino segreto* of Italy, Spain, and France became places in which one might not only seek solitude or recreation, but also conserve and grow interesting, and even exotic, plants. In the sixteenth century, this naturally led folk with orderly minds to begin to collect plants. The first recorded botanical collections were developed in Tuscany, in Padua in 1546 and Pisa in 1547. The Pisa garden was established by Luca Ghini (1490–1556), who was succeeded as director by his pupil, Andrea Cesalpino (1524–1603). Cesalpino was the world's first important botanist, predating Carolus Linnaeus (1707–1778), the Swede who developed the system we use today for the naming of plants and animals. From his garden and collecting trips around Italy, Cesalpino wrote seven books and made a landmark collection of dried plants, comprising 260 pages in thick red leather folios, now in the natural

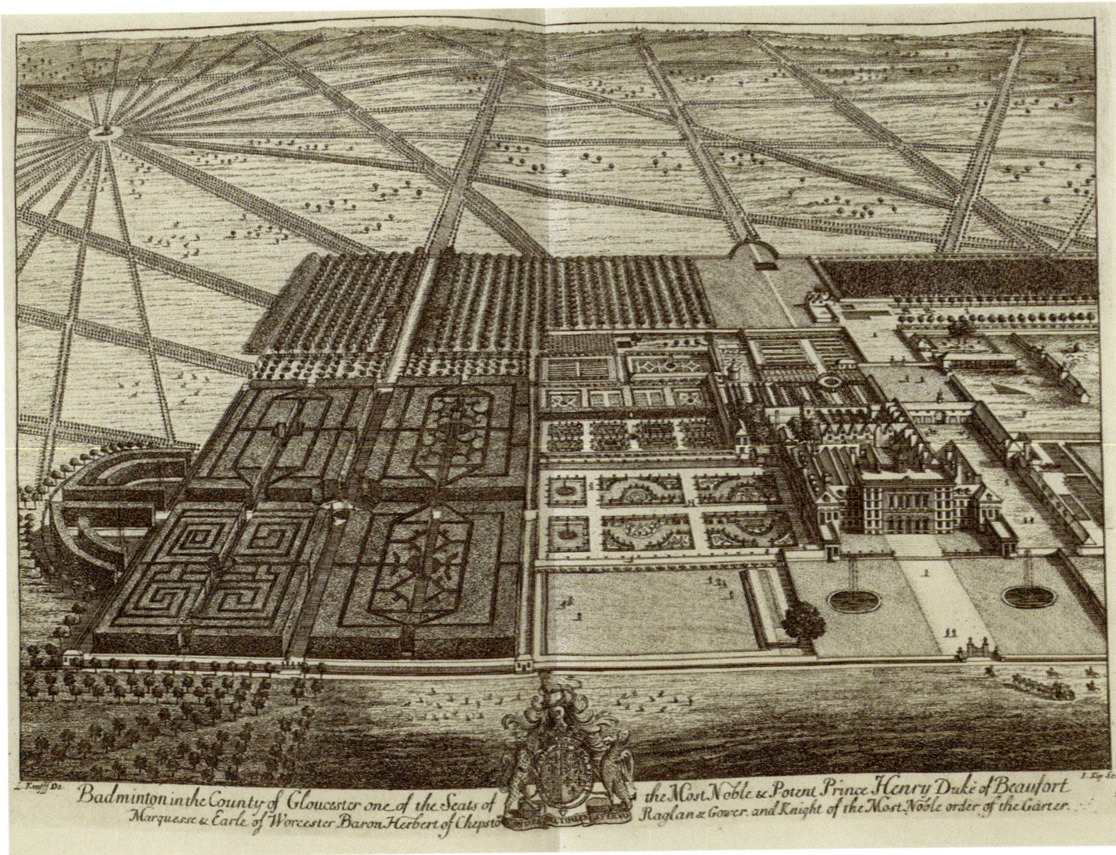

Figure 9
Leonard Knyff (illustrator) and Johannes Kip (engraver), *Badminton in the County of Gloucester, one of the Seats of the Most Noble & Potent Prince Henry, Duke of Beaufort*, 1724 (engraving). The handling of perspective in this line engraving gives a sense of the grandeur of the scheme, guiding the viewer through a series of enclosed spaces, each with its own planting and, in some cases, a water feature.

history museum of the University of Florence. Cesalpino's interest in the agriculturally important pea or legume plants caused Linnaeus to name this plant family Caesalpiniaceae.

In early eighteenth-century Europe, household, park, and landscape design went in two directions. One trend favoured a formal "garden of artifice": a continuation of the formality of Versailles, decorated with statues and flower plantings. The other school reacted against this formality and the separation it created from natural landscapes. Instead, it aimed to create "living paintings" where "naturalness" was recreated through three-dimensional compositional strategies. Underpinning this aesthetic was the notion of artificially creating associations or linkages: a spirit or sense of place more akin to Zen philosophy than the orderliness of the Italian or French style.

The eighteenth-century English picturesque movement was, in part, derived from the visual stimuli of rural subjects portrayed in Dutch prints of a century earlier (Fig. 10). Dutch prints of agrarian and village scenes, including farming folk and animals, depicted a rural way of life not unlike that across the Channel and heightened awareness among the English of their own countryside. English artists also made etchings and engravings after the subjects of Dutch paintings. These prints are among the highest accomplishments of the northern Renaissance, and were widely collected and admired for their pastoral subjects and their high level of technical accomplishment.

As people migrated toward towns and cities, and as the expansion overseas began, the taste for rural memorabilia grew (Fig. 11). Artists from the Low Countries recorded passing features of rural life, while in England the picturesque movement took a last look at the landscape, and at rural life, before the Industrial Revolution altered both forever. Images in this genre are filled with loving attention to endangered ways of life.[12] At the same time, the formality of earlier English landscape design gave way to a new informality whereby nature, rather than a geometric master plan, ruled. This landscaping or "natural" recreation of disorder and romance attempted to reflect contemporary romantic landscape art. Layouts were nongeometric. Villages were moved, rivers diverted to look more pleasing, hills levelled or raised, lakes created, and artificial ruins built. Lancelot "Capability" Brown (1715–1783), to whom we owe our conception of an "English park," created sweeping "landscapes" in this vein. His asymmetrical designs aimed to create "naturalness," an effect he often strove to achieve through major earthworks, shifting mature trees and clearing and replanting land to replace more

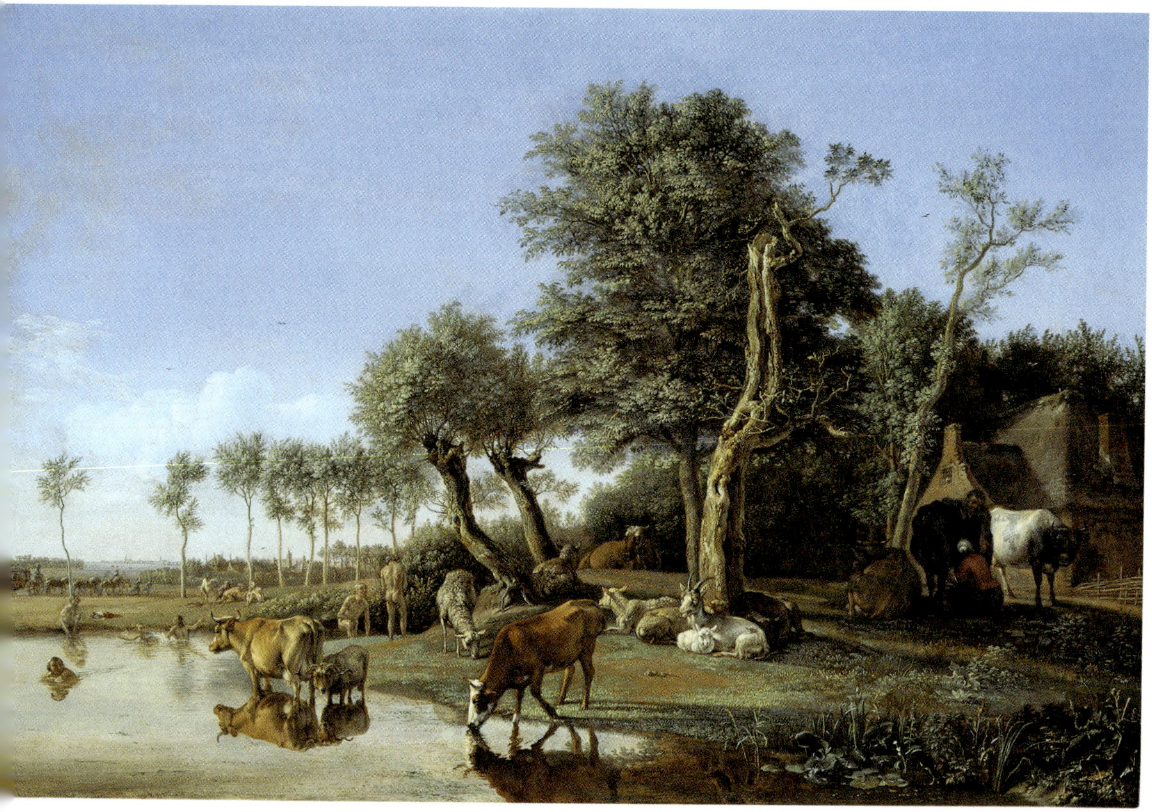

Figure 10
Paulus Potter, *Cows Reflected in the Water*, 1648 (oil on panel). During the Dutch "Golden Age" of the seventeenth century a high level of realism was achieved in the depiction of landscape and animals. In this painting Potter draws on classical sources as he depicts naked cattle herders bathing in a tranquil landscape.

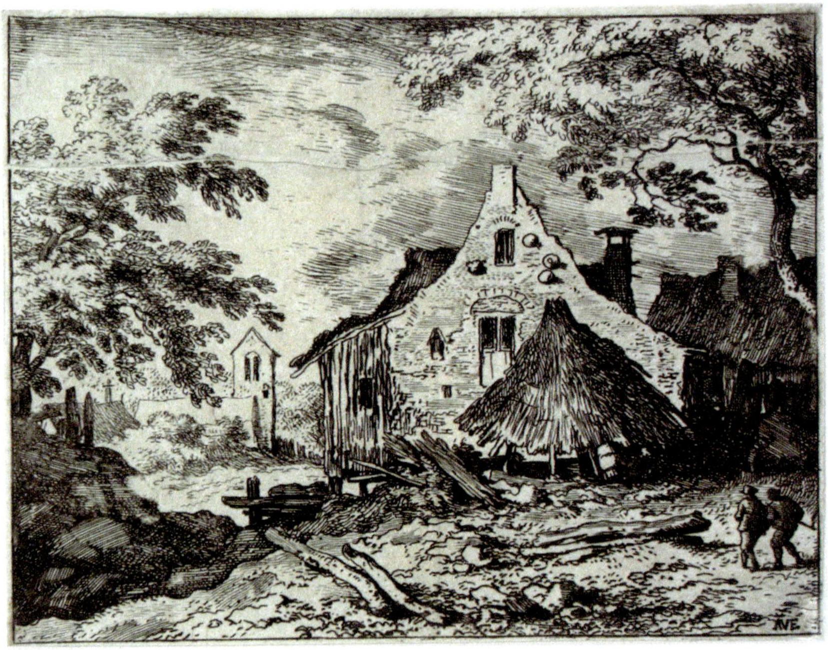

Figure 11
Allart van Everdingen, *The Hayrick with the Movable Roof*, seventeenth century (etching). This work depicts a Dutch farm house with an attached privy that empties into a stream. The hayrick was a portable thatched roof used to cover hay and was adjustable to different heights. This device was used by Dutch settlers in New York State.

formal gardens. Brown, who was most active in the 1760s and 1770s, did things on a grand scale, once planting a hundred thousand trees in a single project. He situated the house or manor in the garden, dispensing with the walled enclosure: his landscapes were venues for outdoor entertainment and reflection.

Blenheim, an hour's drive west of Oxford, is inspired by Brown. As a landscape it appears to be a natural, slightly wooded park, appropriate for, say, riding through or watching horses or cattle graze. The palace sits within it, bare and bold and pink-grey as befits this prize given to a soldier. Its walk or carriageways invite the visitor to leave the house, perhaps to "picnic." (The term is contemporaneous, dating from 1748.) By contrast, the village housing the workers, and the croplands that supported Blenheim, are out of sight of the manor: agriculture has disappeared from daily view, as from the art of the era.

The redesigning of estates by Capability Brown intersected with the history of botany and agriculture for a brief period at Kew, England. Kew, the site of the Royal Botanic Gardens, was developed as a country retreat by King George II and Queen Caroline in the 1720s. A generation later, in 1759, their son's widow, Augusta, founded it as a garden, a repository of plant diversity for the public benefit. A few years later still, in 1766, Capability Brown was commissioned to design and landscape the place; among other things, he removed buildings and caused a valley to be dug out to create his trademark "naturalness." Thus, by 1770 a park-like setting had been created that would develop into a repository for plants – both living, and preserved in a herbarium. New specimens flowed steadily to Kew from throughout Britain's expanding empire and the world.

Humphrey Repton (1752–1818), also British, continued to design parks to achieve the "naturalness" modelled by Brown. However, he reintroduced a demarcation between house and garden on the one hand, and the surrounding "park" on the other. Gardens were "artificial" and contained flowers, while parks, as in Brown's landscape designs, were artificially "natural." Brown, Repton and their followers created a popular craze for "naturalness" that almost every English landowner caught like an infection. Painting reinforced its popularity. J.M.W. Turner (1775–1851) captured the wildness of Sussex woodlands in watercolour and pencil sketches, as did John Constable (1776–1837) when he painted at Arundel in 1833. Constable took his themes from southeast England and, unlike painters before him, painted from sketches made directly from nature. He wrote: "The woods hang from excessive steeps and precipices and the trees are beyond everything beautiful. I never saw such beauty in natural landscape before."[13] Similarly, Richard Payne Knight (1750–1824), an eclectic scholar and proponent of the aesthetic of the "picturesque," provided a demonstration of the contemporary passion for transforming a neoclassical estate design into a naturalistic one in *The Landscape, a Didactic Poem in Three Books* (Fig. 12).

Thus in Europe by about 1800 there were four competing movements or priorities for landscape design and land use. First were formal gardens and villas, capturing the classical style that remained most popular in continental Europe. This, also called the ornamental or villa farm, involved the geometric laying-out of gardens and, in many cases, crop fields. Over-elaborate ornamentation and geometry gave way in the 1780s to more simplified and utilitarian design, but there was nonetheless a place for home, garden, crop, and grazing land. In contrast to these, the park landscape emerged, inspired by Brown, Repton, and contemporary painters, recreating nature's aesthetics for outdoor living, recreation and entertainment. The park landscape was compatible with grazing, but removed agriculture and its labourers from view. Writing in 1825, the reformist politician and agriculturalist William Cobbett (1763–1835) contrasted this design, which he disparagingly likened to a piece of stage scenery, with agriculture and the "sustenance" of a working landscape "of mean farm buildings rented by round-frocked farmers."[14]

The stage is set. By 1700 agriculture in much of Western Europe had become more highly managed by owners and tenant managers, who would benefit directly from innovation and who, no longer dependent on slave or tithed labour, paid for their farm inputs. By 1750 technical innovations had been made that, although not widely represented in contemporary art, would in the nineteenth century change the face of agriculture and land use.

Also at about this time, a sixth child was born to a well-to-do family of gentleman farmers. The biographers of Thomas Robert Malthus (1766–1834) paint him as quiet, genteel, good-natured, and a tad eccentric, which might reflect the fact that he lived mostly with

Opposite: Figure 12
T. Hearne (illustrator) and Benjamin Thomas Pouncy (engraver), Illustrations No. 1 and No. 2 from Richard Payne Knight's *The Landscape, a Didactic Poem in Three Books*, 1794 (etching). This before-and-after proposal shows in the lower image the existing serpentine stream, simple arched bridge and Palladian house. The upper register shows the proposed improvement: a romanticized Elizabethan-style mansion in an overgrown picturesque setting complete with a quaint modern bridge.

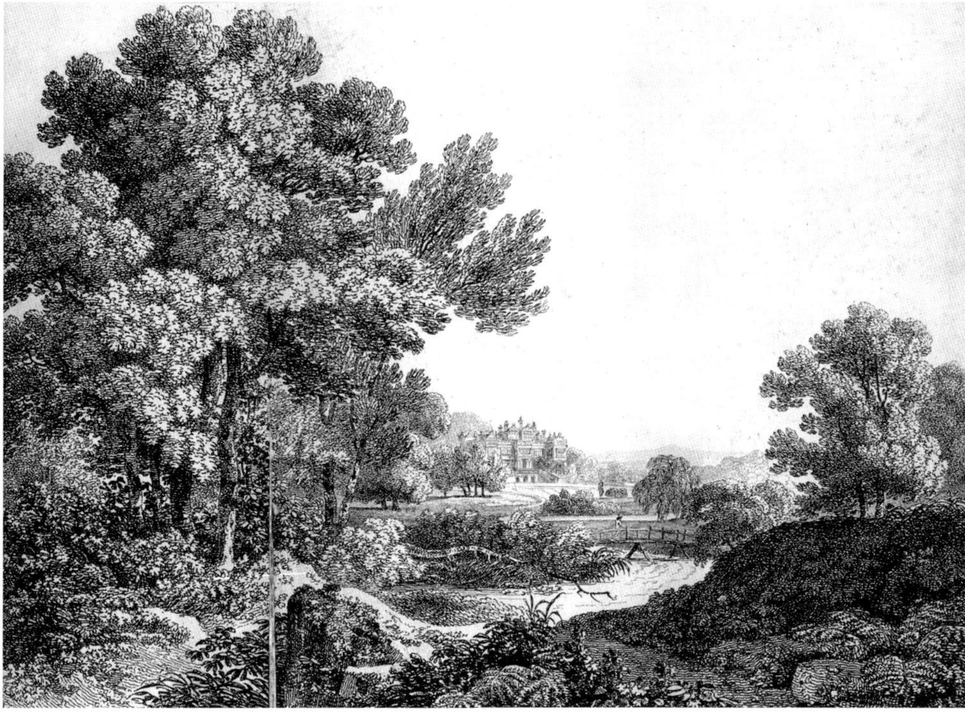

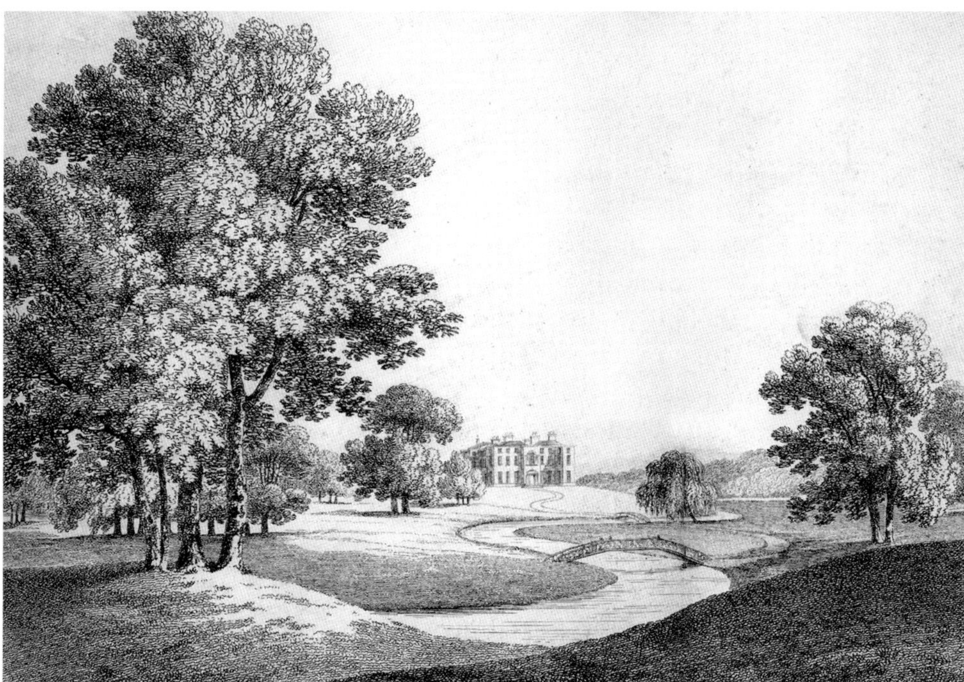

his parents until they died in 1800. He studied mathematics at Cambridge and agonized over what was, at least for him, a crucial question: If humanity is defined by a sense of personal responsibility, as he strongly believed, then would social welfare encourage, or work against, the building of this sense of personal responsibility? At a time when Britain's population had grown rapidly to over ten million (although he thought it about seven) he proceeded, in his late thirties, to pour a bucket of cold water on the prevalent view of society, fostered by the recent revolution in France, that egalitarianism would lead to a utopian future. His *Essay on the Principle of Population* was published anonymously in 1798 and signed in the second edition three years later. It started with two premises: "(1) That food is necessary to the existence of man; and (2) That the passion between the sexes is necessary, and will remain nearly in its present state." From this starting point Malthus proposed that "the power of population is definitely greater than the power in the earth to produce subsistence for man. Population when unchecked increases in a geometrical ratio. Subsistence increases only in an arithmetical ratio. A slight acquaintance with numbers will show the immensity of the first power in comparison with the second."[15]

In about fifty years, Mathus predicted, there would be famine and vice. Only famine, pestilence and war would check population from outstripping food supply. Malthus' bleak view of the future gave rise to the saying that economics is "the dismal science." He encouraged this view himself, in his anonymous preface to his first edition: "The view that [the author] has given of human life has a melancholy hue." Dismal or realistic, later economists, most notably John Maynard Keynes (1883–1946), would call Malthus a "youthful genius." His view of the race of agricultural innovation against population growth remains the centrepiece for debate about food security over two hundred years later. Thus, as the manorial era closed in the early 1800s there were widespread utopian views of a sunshine-filled agrarian future, "dismal" voices presaging famine and pestilence, and new views of landscape design that sought to create idealized "nature." This aesthetic would in the next century give rise to public parks and wildlife reserves, the first of which was created by 1872 at Yellowstone National Park in the northwestern United States. In the meantime, migration out of Europe and the possibility of a vast expansion of farming by a wave of settlers in the New World had begun to give rise to a new paradigm of agriculture – the subject of our next chapter.

2

NEW WORLD EXPANSION

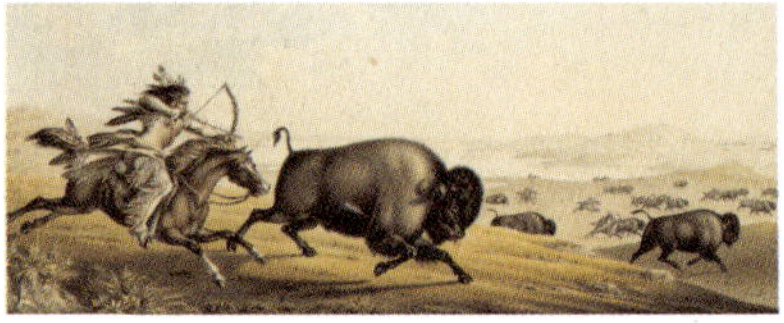

While grand manors and enclosed fields remained the centrepieces of agricultural and other planned landscapes in Europe, new landscapes were being fashioned in tandem with European migration to the northern and southern hemispheres. Western agriculture's age of expansion began in about 1600, gaining momentum over the next few hundred years with the relentless clearing of land and increasing mechanization (Fig. 13), and peaked in the second half of the twentieth century – from about 1950 to 1970 in the United States and Canada, between 1975 and 1985 in Australia, Argentina and New Zealand, and in the 1990s in Brazil. In all cases, an expansionist paradigm emerged much earlier than its full realization (Appendix 1).

To graduate in agricultural science in about 1970, in Western Australia, was to experience the late, and most dramatic, phase of the age of expansion. There is something very satisfying about driving a header ten metres wide, perched high above the waving grain as it remorselessly pushes through the field at harvest, blowing a plume of chaff vertically into the deep-blue sky and the shishing grain into a bin at the side, ready to be off-loaded into a field-bin or

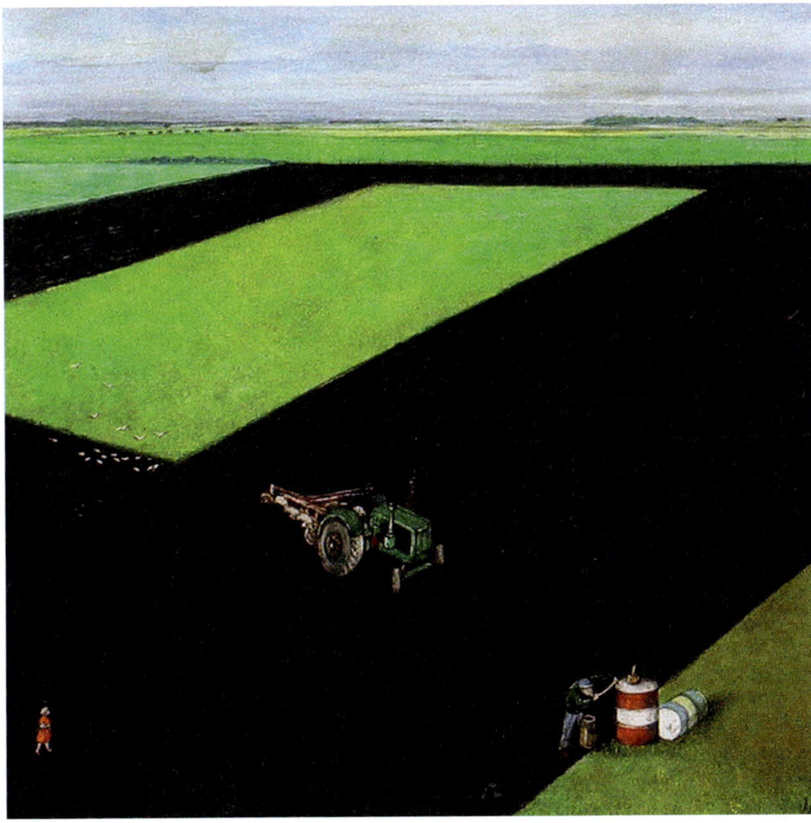

Figure 13
William Kurelek, *Plowing*, not dated (acrylic, graphite, coloured pencil on masonite). Kurelek captures the grandeur of the wheat harvest on the Canadian prairie, creating an image of the sublime that recalls Constable's views of Hampstead Heath. Kurelek both romanticizes and acknowledges the farmer's role as essential to human survival.

truck and transported to the railhead. Round and round the field one goes, making a clear cut against the previous pass, until only forty-centimetre stalks stand stiffly across the flat landscape. Rooted in the hot sand, chewed and trampled by cattle or sheep, they give the landscape a brown, bristled appearance until the autumn weather breaks, the soil becomes wet, and the cycle begins again.

In Western Australia, expansionism peaked dramatically in 1970 with the clearing of three million acres of land. A frontier mentality prevailed. Land was plentiful, released from the government for sale at low prices provided the purchaser settled it, and settlement was made possible by cheap methods of clearing the native vegetation. Although this expansion appealed to a sense of pioneering adventure and entrepreneurship, it relied on the good prices for agricultural commodities (notably wheat) created by demand from the rapidly expanding populations of Asia and Africa. The resources of the uncultivated landscape – forests, minerals, and the biodiversity of plants and animals – were irrevocably altered in the service of providing food and materials for the Old World and creating local wealth in new colonies.

Although the idea of "agricultural expansion" might imply a stage in a linear process of development, it was not so. The paradigms of expansion, production, productivity, and sustainability overlap to this day. An example is the clearing of native vegetation. Land-clearing in Australia continues at about 500,000 hectares a year, despite the fact that most of the population now support the preservation of native vegetation and biodiversity, and despite growing concern with environmental issues such as climate change. However, to consider these overlapping paradigms in sequence, let us go back for a moment to the seventeenth century.

Europe was settled. Agriculture was stable. Communities were tightly knit, and there were close ties between most people and the farms or garden plots from which they obtained their food. Then followed an unprecedented period of settlement as colonists succeeded in finding good soil in foreign continents. The motivations for colonization were wide-ranging. They included expansionist politics and a desire for a source of cheap food, straight timber for ships' masts, and other materials. Colonies also created a conveniently distant repository for criminals and a refuge for those escaping persecution for religious beliefs.

Early colonies, whether in Atlantic North America or eastern Australia, clung to bays and alluvial flats. Settlements were surrounded by seemingly impenetrable native vegetation and by hunter–gatherer societies – both of which were in different ways feared, fought and romanticized by the European settlers. Samuel de Champlain's misguided attempt to bring European horticultural practice

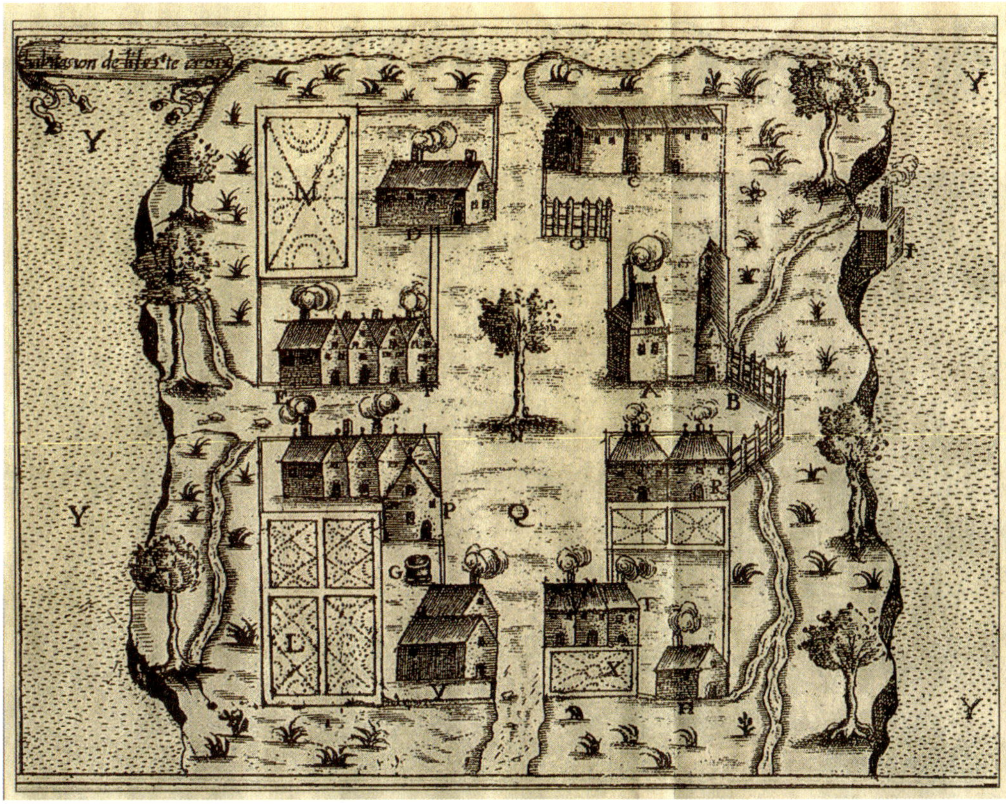

Figure 14
Samuel de Champlain, *A Plan for the proposed settlement at Isle Saint-Croix on the south shore of the Bay of Fundy*, 1613 (ink sketch). The French explorer published this settlement plan in his travel journal to document the failed 1604 settlement on Isle Saint-Croix in the Bay of Fundy. The plan shows vegetable gardens laid out in parterres following the European style, which mimicked Versailles – an inappropriate scheme for the rigours of a Canadian winter. The settlement failed within the year, but Champlain noted that settlers who returned the following year found lettuce, sorrel and cabbage growing among the weeds.[1]

to the wilds of North America (Fig. 14) is only one example of the clash of cultures between the colonists and native peoples. So intent were the settlers on imposing their notion of civilization on North American aboriginal people that they failed to notice such highly successful native agricultural practices as harvesting wild rice (Fig. 15). The Iroquoian people of the St. Lawrence Valley and Great

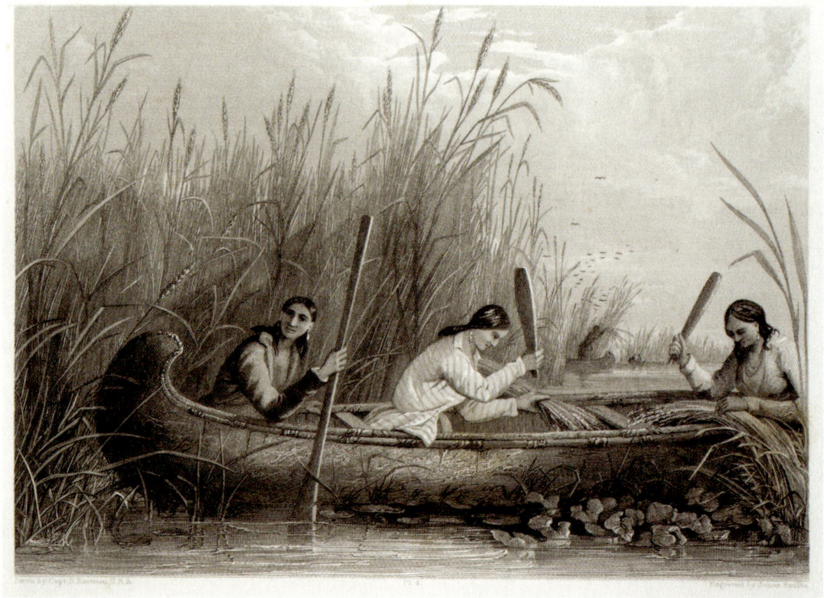

Figure 15
Capt. S. Eastman (illustrator) and James Smillie (engraver), *Gathering Wild Rice,* in Henry Rowe Schoolcraft's *Historical and Statistical Information, Respecting the History, Condition and Prospects of the Indians of the United States*. Philadelphia, 1851–57, Vol. 3, Plate 4. An Indian agent with headquarters in Sault Ste. Marie in 1822, Schoolcraft published his major work in six volumes from 1851 to 1857; this was one of the first publications on American native civilizations.[2]

Lakes region were successful farmers long before European contact. By the seventeenth century, they were growing approximately 150 different varieties of corn. It has been estimated that over 500 plants were used by Amerindians for food or food preparation, and another 500 for medicinal or ritual purposes.[3] Instead of sharing the rich alluvial farmlands with the native growers of maize, the settlers practised a systematic genocide, driving indigenous peoples farther and farther west to mountainous and arid lands unsuitable for the agricultural methods of the eastern tribes.

The settlement of North America was a multi-pronged affair, the French making incursions from the northeast through the St. Lawrence Valley and from the south up the Mississippi; the Spanish arriving from the southeast, settling Florida and expanding across the Mississippi, and moving up from the Caribbean through Mexico and into what was to become Texas, Arizona, and southern California. Russians came from the northwest, overlapping their forts and trading centres with the missions of the Spanish, Dutch, Swedish, and English.

Settlers were impressed by the vegetation they found in their new territories. Along the northeast coast was an abundance of trees for firewood – a contrast to old Europe, where wood was scarce and expensive. Early settlers writing to friends in Europe listed the various plants and animals they found and protested to their correspondents that, although their reports might be hard to believe, they did not exaggerate the richness of the flora. Settlers wrote that at Plymouth there was "good ground in abundance, with excellent good timber" and "Though it bee here somewhat cold in the winter, yet here we have plenty of fire to warm us, and that a great deale cheaper than they sel billets and faggots in London."[4] Some, visiting Narragansett Bay in 1524 and Massachusetts Bay a century later, described open woodland; others, north and inland, found multi-storeyed temperate forests. Settlements crept along the east coast and river estuaries. The colonists were toughened by the seasonality and hemmed in by poor harvests and a lack of transport for anything larger than a person on a horse.

John Howison (1797–1859), an Edinburgh-born doctor, writer, and traveller who spent two and a half years in Upper Canada, reflected on travel conditions in about 1820:

> Often, for some weeks in winter, the sky continues bright and cloudless, and though the air is intensely keen, yet its bracing and exhilarating effects enable one to bear the cold without any inconvenience. At these times the public roads are crowded with sleighs … . Spring commences in March but the early part of the season is seldom agreeable, being damp, tempestuous, and rainy. The roads likewise become so bad, that it is hardly possible to go out of doors.[5]

Despite their foreignness the early settled areas grew, and within them a new aristocracy emerged that, through trade with Europe, local commerce, and agriculture became rich and built grand houses in the tradition of the countries they had left behind: the houses and lifestyle of these leading colonists, grounded in agriculture, has been called the *belle époche*. Manor houses dominated the landscapes: Monticello in Virginia (constructed from 1769–84, remodeled from 1796–1809); Elizabeth Farm, a half-day carriage ride west of Sydney (1793); or Talgai Homestead (1868), a stunning manor on grazing land three days' ride from Brisbane. In Argentina, *estancias* (colonial estates) were established near the coast along the Paraná and La Plata rivers beginning in the 1500s and the first half of the 1600s. On these grand farms, some of which are still intact, the main house served as the centre of the business establishment. Small, elegant gardens with water features and baroque sculptures created tranquil buffers between the manor house and the vast farmlands. The use of the baroque style was a nostalgic reminder of the grand estates in Spain and a symbol of the colonists' success in the New World.

Perhaps the most famous of these grand manors in the United States is Monticello (Figs. 16, 17), a Palladian villa designed by Thomas Jefferson (1743–1826). The chief author of the American Declaration of Independence, Jefferson died on the fiftieth anniversary of its adoption. Within that half-century he built and remodelled his house, recording his plans and trials in his voluminous diaries and correspondence. Jefferson read about European gardens and, in 1786, while Minister to France under Washington, visited English gardens recommended in contemporary books. However, like the grand manors in South America and Australia, Monticello is not simply a copy of the manor houses of old Europe. As one garden historian writes, "As far as one can judge, what Jefferson did at Monticello was without precedent in [North] America, nor was it any literal adaptation of what he knew about the latest European gardens."[6] The villa dominates the site and, because it is on a hilltop, suffered from periodic shortages of water. Jefferson also designed "roundabouts," ring roads that circled up the hill for easier ascent and to afford pleasant views. He wrote to his granddaughter: "I have resumed an idea for a winding walk surrounding the lawn before the house, with a narrow border of flowers on each side."[7] He is credited as the first to design grand gardens with flowers in the New World.

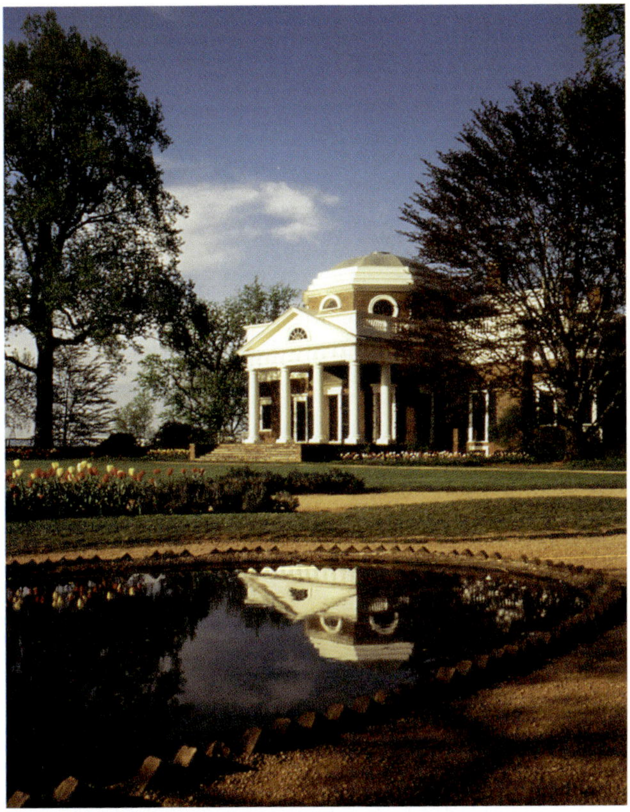

Figure 16
The fish pond at Monticello, with a view of the west front of Jefferson's Palladian villa.

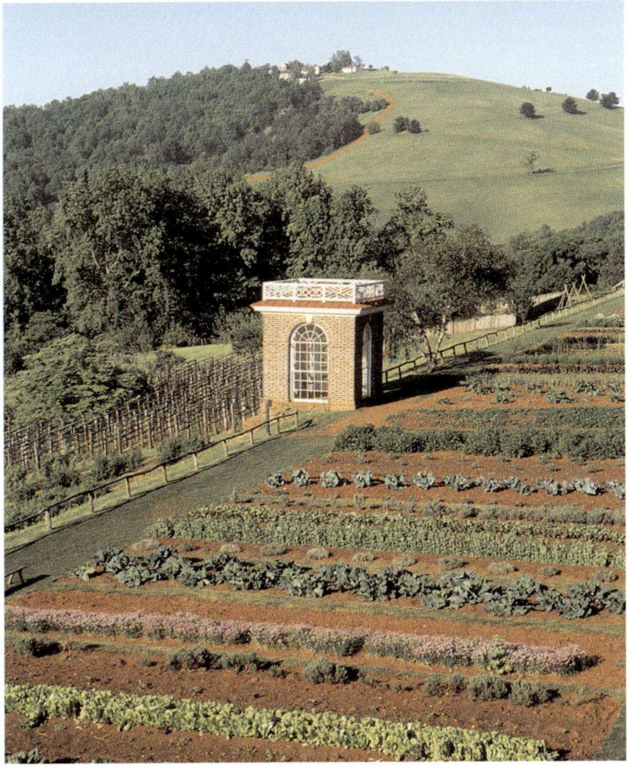

Figure 17
The vegetable garden at Monticello has been restored with early vegetable varieties.

Much as Jefferson admired the style of European gardens, the aridity of Virginia led him also to emphasize trees: "Trim up their bodies as high as the tree will bear, but so that their tops shall still unite and yield dense shade."[8] As in Europe, trees were planted on the Monticello estate in rigid symmetrical patterns; their sites, which it has been possible to recover, have been found to correspond exactly with the plans of the time. A vineyard – it must have been one of the colonies' earliest – was planted at Monticello in 1770, but met with a succession of failures. Other fruit crops fared little better: the peach orchard was completely killed in 1775, the cherries and peaches (again) in 1806. Like most of the grand estates built near the east coast in the early days of New World expansion, Monticello fell on hard times. It was sold in 1831 to pay off the debts of the Jefferson estate and decayed for nearly a century as wealth moved from its original agricultural base near the first settlements, into the growing cities or to the west, where it could be sustained through large-scale agriculture.

Away from the coast, explorers, trappers, and traders followed river routes and trails established by native Americans to gain access to the interior. A woodcut of an animal suggesting a bison (Fig. 18) was published in a book printed in Antwerp in 1557 intended for European readers. Many illustrations in books about expeditions to the New World were done by artists who actually never visited but relied on the enthusiastic descriptions of explorers. The bison looks more like a curly haired dog with hooves, a lion's face, and horns.

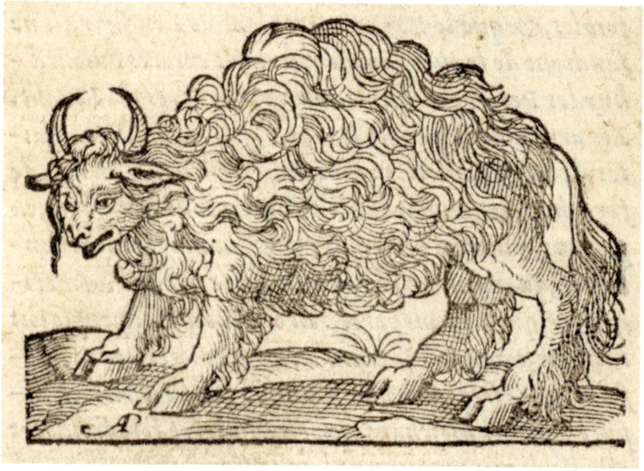

Figure 18
André Thevet, *Strange Beast* (from *Les singularitéz de la France Antarctique*), Paris 1557. This woodcut appeared in a book published for a European readership. Many accounts of expeditions to the New World were illustrated by artists who had never visited the New World but relied on the enthusiastic descriptions of explorers.

The woodcut does indicate, however, that the short-grass prairies were known and recognized to support large game.

David Thompson (1770–1857), an English-born Canadian explorer, criss-crossed the interior by foot and canoe, travelling fifty thousand miles over twenty-seven years to trade furs and survey lands for the Hudson's Bay Company. Thompson was the first European to describe the Great Plains as: "lands bearing grass, but too short for the scythe ... These Great Plains may be said to commence at the north side of the Gulf of Mexico, and extend northward to the latitude of fifty four degrees; where these plains are bounded by forests of the north."[9]

Travellers were also eager to document the exotic lives of the American Indians. Wealthy Europeans, such as the Czech Volta Nápresk, travelled among aboriginal groups in America to acquire the best possible examples of their hunting, domestic, and spiritual artifacts. In 1856, Little Crow, chief of a group within the Dakota Mdewakantons, presented Nápresk with a woman's robe made of bison skin and decorated with an image of a bison. The robe became the first American object for the museum of world cultures and industries that Nápresk established in Prague in 1862.[10]

A fascination with aboriginal cultures extended to European-trained artists, who created numerous picturesque views for society drawing-rooms (Figs. 19, 20). Although these watercolour and oil paintings were not always accurate, they did provide some insight into native culture.

It is worth noting that "many supposedly wild landscapes inhabited by hunter–gatherers – the North American prairies and Australian outback, for instance – were shaped by deliberate fire-setting" by Aboriginal peoples to extend the grazing lands for animals such as bison and kangaroo.[11]

As European settlement pushed eastern tribes westward, there was enormous pressure on native peoples' traditional way of life and their view of the land as sacred. Their lives were adversely affected by warfare with the United States cavalry, introduced diseases (notably smallpox), Christianization, forced removals, the provision of inadequate and inappropriate agricultural tools, residential schools,

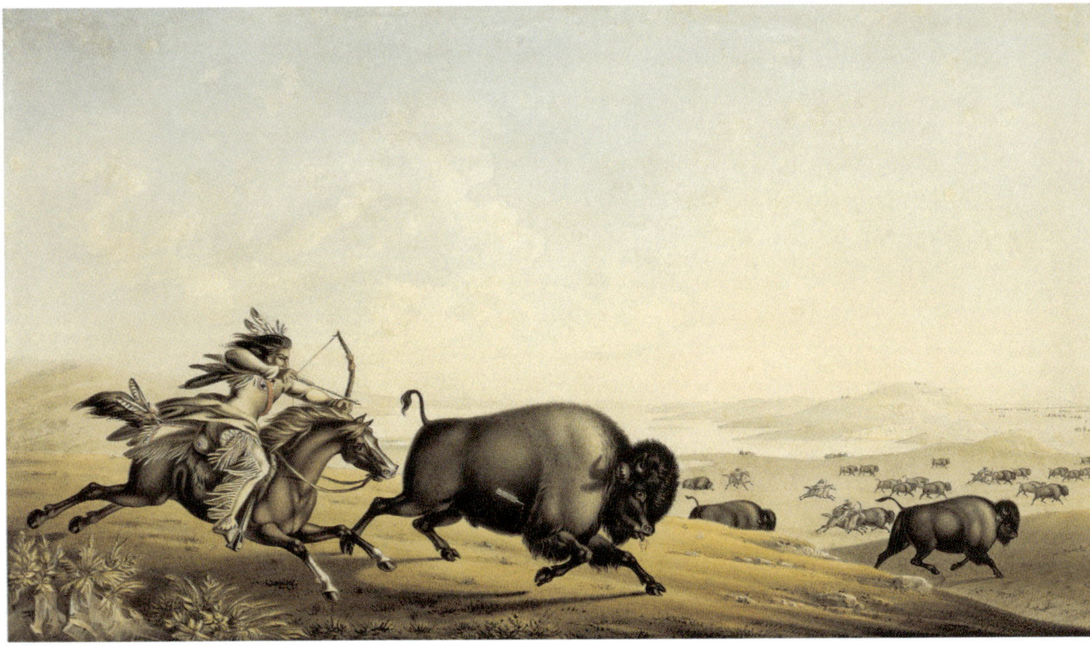

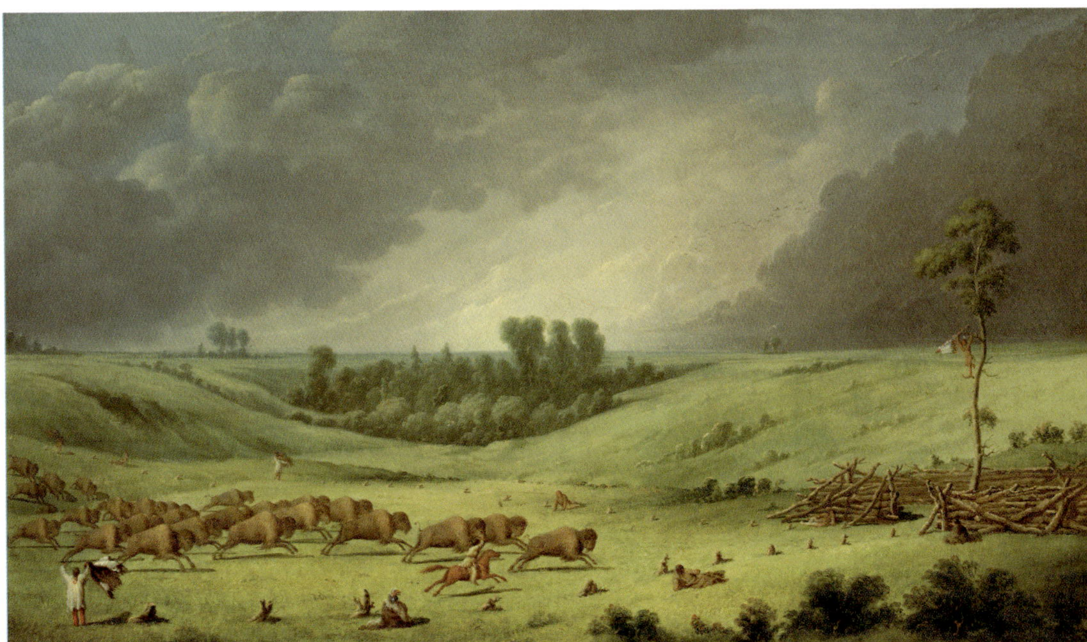

and the availability of alcohol. The buffalo slaughter that artist/explorer Paul Kane witnessed and documented (Fig. 20) might be construed as an act of desperation in the waning days of the archetypal hunter–warrior. The natural-conservation approach to nature, which required the hunter to take only what was needed and leave some for the next generation, was shattered. Despite some prescient observers who foresaw the results of over-hunting, in the nineteenth century the belief prevailed that bison and other plentiful species were an inexhaustible resource. Euroamerican travellers engaged in mass slaughter of the buffalo as sport, aided by the passenger trains that now rolled across the prairies, providing comfortable seats to shoot from.

Before long, the days of the boundless hunt gave way to the management of livestock. Kane observed Métis herding cattle in or near the Red River in Manitoba – an indication of the shift from bison hunting to cattle grazing. In 1847, Kane pushed farther west to Fort Vancouver and then south to Oregon City. On April 7, he arrived at Fort Nisqually, the headquarters of the Puget's Sound Agriculture Company, a subsidiary of the Hudson's Bay Company that was established in 1840 to promote settlement of the Oregon

Opposite top: Figure 19
Peter Rindisbacher, *Assiniboine Hunting on Horseback*, 1833 (watercolour with glazing over graphite underdrawing on paper). Rindisbacher (1806–1834) was a Swiss-born artist who, although he lived to only age twenty-eight, created realistic views of native life that were distributed through periodicals and lithographic reproductions. This watercolour depicts the dramatic pursuit of a stampeding buffalo herd by hunters using bows and arrows.

Figure 20
Paul Kane, *A Buffalo Pound*, 1846 (oil on canvas). Kane (1810–1871) was an Irish-Canadian artist also known for recording in sketch format the First Nations peoples. On the basis of a field sketch, although with altered details, Kane created an oil painting that depicts the Indian hunting technique of herding bison into a walled compound built with logs and stumps. As soon as the herd was in the enclosure, the entrance would be closed. The buffalo were then killed with bows, arrows and spears. Kane describes a buffalo pound he observed on September 12, 1846, about six miles out of Fort Carlton, which he visited on a journey from Fort Carlton to Fort Edmonton: "Whilst the buffaloes were being driven in, the scene was certainly exciting and picturesque, but the slaughter in the enclosure was more painful than pleasing. This has been the third herd that had been driven into this compound within the last ten or twelve days and the putrefying carcasses tainted the air all around."[12]

region. In his journal he comments on the large quantities of wheat being grown at Cowlitz Farm, a short distance to the south: "When I visited it, it had about 6,000 sheep and 2,000 horned cattle. Its site is beautiful on the banks of the eastern end of Puget's Sound. The land is inferior to that in some other parts of the same district, the soil being gravelly; the grass, however, grows luxuriantly, and the mildness of the climate adapts it well for grazing purposes, as it is never necessary to house the animals. The wool, which is good, finds its way to English markets by the Company's ships, and the cattle are slaughtered and salted for the Sandwich Islands and the Russian Dominions."[13]

Kane's eyewitness report provides an example of an integrated production and marketing enterprise operating on the coast with easy access to shipping. At the same time, an industry based on fur trapping had developed. The Hudson's Bay Company and similar trading enterprises bartered with aboriginal people, taking their furs in exchange for items such as tea, sugar, flour, cloth, and ammunition. The fur market continued until World War II, when fashions changed and there was no longer a demand for white-fox furs.

The Scottish-born fur trader Alexander Mackenzie (1764–1820) travelled in 1789 from Fort Chipewyan on Lake Athabaska northwest along the Mackenzie River to the Arctic Ocean and then, four years later, overland from the Fraser on to the Pacific Ocean, arriving at the mouth of the Bella Coola River. He wrote: "The whole of this country will long continue in the possession of its present inhabitants, as they will remain contented with the produce of its woods and waters for their support, leaving the earth, from various causes, in its virgin state. The proportion of it that is fit for cultivation is very small, and is still less in the interior parts; it is also very difficult of access."[14]

However, other travellers were aware that the West was a source of curious and potentially useful plants. In their expedition to the Pacific of 1803–06, sponsored by Thomas Jefferson, Meriwether Lewis and William Clark collected many plants that would mark their expedition as worthy, in their eyes, in the historical company of explorers and collectors such as James Cook.[15] During their two-and-a-half-year journey, the settlement of the West began to accelerate – they evidently met as many as 150 settlers pushing west as they returned to St. Louis. Those plants most likely to be pleasing in gardens were planted by Jefferson (he was president at the time), or sent to Kew, and grown in European gardens.

The first commitment to the permanent occupation of western North America by Europeans was made in 1796 when General Moses Cleaveland formed a settlement that now bears his name, albeit with an altered spelling. However, the big step toward inland settlement across the Appalachians was to wait until the convergence of a number of factors, both psychological and economic, after the conclusion of the War of 1812–14 between the United States and Britain. At the end of the war, the United States government provided land grants for veterans. The government also built a National Road from Cumberland, Maryland to Wheeling in what is now West Virginia. By 1821, over a million people had moved west of the Appalachians, most notably into the closest area, Ohio, where 600 thousand settled. Settlers spread across Ohio and took land on what are today the twelve states of the cornbelt. The historian Frederick Jackson Turner (1861–1932) summed up western expansion this way, "The exploitation of the beasts took hunter and trader to the west, the exploitation of the grasses took the rancher west, and the exploitation of the virgin soil of the river valleys and prairies attracted the farmer."[16]

Contemporary accounts took a mixed view of these developments. On the one hand, land west of the Ohio and Mississippi was advertised to potential settlers from Europe as a land of promise. On the other, contemporary novelists played up the hardships: the need to conquer nature and a climate that, depending on the season, was so hot that "not even an insect crawls about" or so cold and snowy that wagon trains were blown off course, lost, or perished.[17]

Turner, who taught at the University of Wisconsin and at Harvard, promulgated his highly influential "frontier thesis," according to which, by taming nature, individuals contributed to a "new order" and shaped American values of independence and democracy. Tragically, this new order had no space for Native Americans: General Custer's massacre of 160 Cheyennes at Sand Creek twelve years before the Battle of the Little Bighorn is one of many atrocities. The surviving tribes, some of which were reduced to a tenth of their previous populations, were relegated to reserves (in the United States) or reservations (in Canada) on less fertile or inaccessible land. Nor was much thought given to the landscape: "Reflection and well-considered planning seem wasted time."[18]

Although steamboats along the Ohio and Mississippi rivers allowed crops to be transported relatively cheaply to markets, it was a long haul from Cleveland to New York via the Gulf of Mexico. In 1825 the Erie Canal was opened, connecting Buffalo on Lake Erie to Albany on the Hudson River. From Albany, goods could be transported by barge or ship to New York, or by train due east to Boston. The city of Buffalo began to grow, and New York became a major trading centre. Its interesting to speculate what might have happened to New York had the Erie Canal not been dug: the Canadians were advocating transport to the east coast and Europe more cheaply along the St. Lawrence, which had locks to Montreal by this time.[19] The use of the Erie Canal gave the cereal-growing areas west of the Ohio access to east-coast settlements and markets. These settlements used about 1.5 million barrels of flour in the mid-1800s. Assuming that only 600 barrels could be carried on each horse-drawn canal barge, the traffic on the Erie Canal would have exceeded ten barges a day. The continuous plodding of draft-horses on the tow-paths along the banks of the Erie Canal contrasts with the tranquility of the waterway today, broken only by landing water-birds and the once-a-day weekend sailor.

Other canals followed as each town recognized that growth depended on the cheap transport of inputs to the developing west, and of grain and cattle to markets in the east. By 1830, agriculture had changed the landscape in the eastern half of the continent, both directly and through the creation of a network of roads and canals, while west of the Mississippi still lay a land of trade routes and trading posts.

In 1860–61 the Pony Express linked the continent from the east at St. Joseph on the Missouri River near Leavenworth, to Salt Lake City, to Sacramento and the sea. Settlement followed its route. Turner commented that in 1880 the US "had a frontier of settlement but that by 1890 it was impossible to draw a line between the settled and unsettled areas, that the westward advance of the frontier had

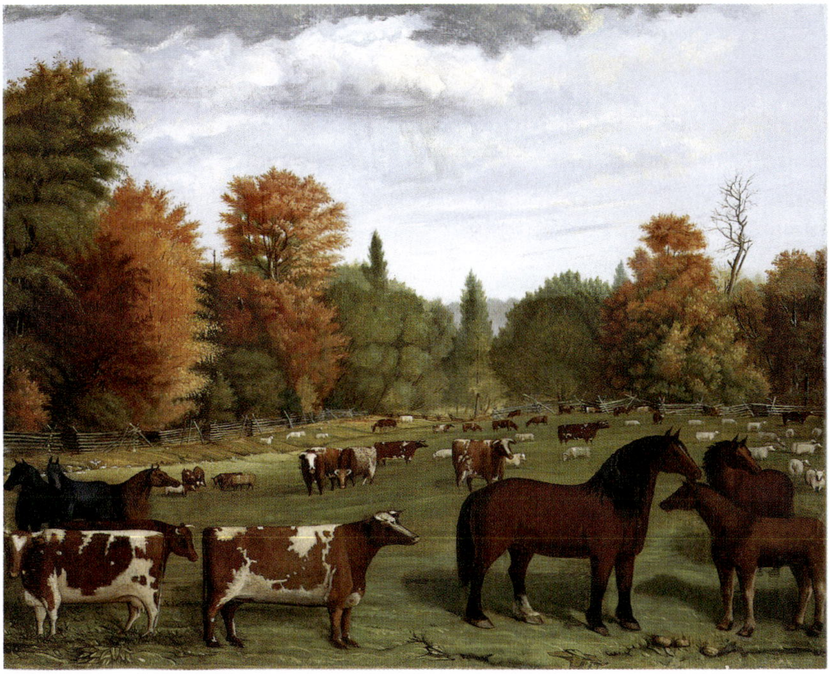

Figure 21
Ebenezer Birrell, *Good Friends*, circa 1830 (oil on canvas). Birrell likely painted this scene on his own farm near Toronto. Cattle, horses, and sheep graze in a peaceable kingdom. The tranquility of the scene and the healthy, contented livestock are meant to convey his gratitude for the plentitude of the land and nature.

ended."[20] Turner, among others, hypothesized that, during this most rapid age of expansion it was the frontier – the unexploited land and not the settlements themselves, that provided the colonists' mental focus and the imperative to expand.

In parallel with the expansion to the west, agriculture intensified in the areas of older settlement closer to the coast. This led one observer, writing in 1821, to describe upper New York State in the following terms: "The country, a little way beyond Utica, is in a high state of cultivation."[21] Here, farms were smaller but managed to survive by virtue of their proximity to markets (Fig. 21). At the same time, that proximity to markets allowed growers to diversify their crops and to shift to exotic varieties and perishables such as fruit for consumption in nearby towns and cities. Schemes to introduce new varieties abounded, as in this suggestion dating from about 1800 in Ontario: "It is believed that the inhabitants of Canada would derive great wealth from the ginseng that grows in swamps, for in China it was worth its weight in silver."[22] East of the Ohio River and close to the Great Lakes, intensive, exotic crops were tried, and tried again throughout the nineteenth century as small industries flowered, flourished, and floundered. Small farmers and entrepreneurs were encouraged by the relative abundance of a wide range of naturally occurring exotic plants: "Native wild grapes, some of excellent form and quality, grow everywhere throughout these [Ontario] counties, and on the St. Lawrence islands on the dry limestone gravelly hills and ridges, the soil being deep and strong ... Many attempts have been made to plant the European grapes ... but in the end they will all succumb to mildew ... they always would prove a failure."[23] Ironically, native North American grapes such as the Concord variety, although often maligned as poor for wine-making, actually saved the French wine industry from the ravages of phylloxera. Phylloxera is a root-feeding aphid-like insect first noticed in France in 1863. By the late 1880s it was threatening the entire wine industry in France and neighbouring countries. The solution was to graft European species onto American rootstocks, which were naturally resistant to the pest. This method continues as the established procedure today for controlling phylloxera.[24]

In the southern United States, some of the settlers in the Carolinas came from the Caribbean, bringing their slaves with them. The period from just before the confederation of the states until the 1850s saw an exponential increase in the number of enslaved people in the United States. They were mostly West Africans who had first been trafficked to the Caribbean. It was the slaves, rather than their masters of European origin, who brought knowledge of wet rice production to the Carolinas – along with the free labour that made it economically feasible to reshape the landscape and engineer the water supply. Planters placed orders for slaves to be acquired from specific regions in Africa where they knew rice was cultivated, so that they would bring expertise as well as muscle-power. The transformation of the southern landscape from swamp to lands that could support one of the most intensive forms of agriculture was a massive achievement:

To understand the magnitude of technical achievement, one need only compare the endless fields of rice in 1840 [in South Carolina] to the virgin swamp forest the slaves had begun to clear by the end of the seventeenth century. ... To create such [rice] fields, the slaves had to fell hard-wood trees as large as ten feet in diameter, remove the timber, dig out the stumps, level the land, and then build the dykes, trenches, and gates that allowed them to control the water level. They did it all with simple hand tools, standing knee-deep in mud and water with the added stress of alligators and poisonous snakes, not to mention extremely high temperatures and humidity. Developing a typical rice field that covered between forty and seventy acres could take up to ten years.[25]

Interestingly, the management of enslaved workers in rice-based agriculture differed from that in the cotton fields. In the case of rice cultivation, they were assigned a task, an amount of work – perhaps ground preparation, or irrigation, or harvest, depending on the time of year. Once the day's task was completed, they could work as they wished, perhaps in a household garden, in a manner not unlike the feudal system of medieval Europe. In the cotton fields, by contrast, enslaved workers toiled in gangs from dawn to dusk. The outcome, however, was the same: a long period of intensive agriculture based on gruelling manual labour that created and maintained the slave-owners as the wealthy social elite. The intensive production of rice and cotton continued to build momentum until, with the Civil War, the land was laid waste: Carolina gold rice and Mississippi cotton gave way to abandoned fields, yellow primrose, and cattails.

In time, the balance of agriculture's shift to the interior was complete, and large-scale commodity production of grains and beef was economical only in those regions, thanks in equal part to appropriate farm size, mechanization, and a government-supported infrastructure of canals and roads to support cheap transport. The towns in the east and north persisted, but the acreage under cultivation shrank. By the early 1900s, valleys were returning gradually but perceptibly to an uncultivated state. Nowadays, along the Erie Canal and through upper New York State, neat white-washed houses (Fig. 22) still cluster together near wood-frame churches, but the landscape has reverted to secondary forests: glorious golds in autumn, but not the fields of corn of an agricultural landscape in "a high state of cultivation."

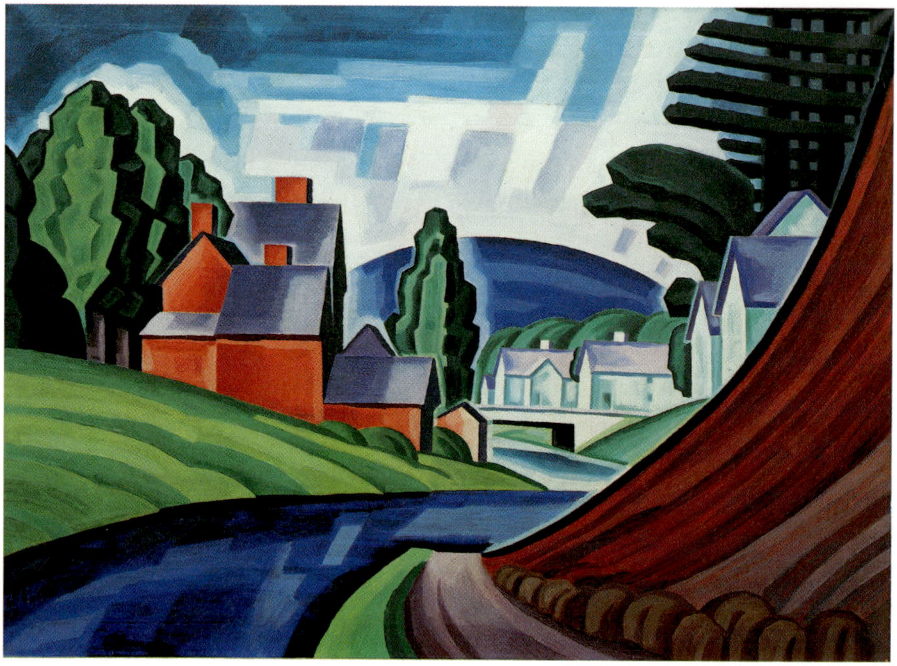

Figure 22
Oscar Bluemner, *Space Motive, a New Jersey Village*, circa 1917–18 (oil on canvas). German architect Oscar Bluemner came to the United States in 1892, settling in New Jersey; he emerged as an innovative painter in the colourist tradition in 1912. Bluemner's work combines the German Expressionist aesthetic with the simplicity of small-town America. Bluemner depicted an "intimate landscape ... where town and country mingle," which he rendered in bold colour and scale.[26]

Prairie settlement suffered most from a shortage of labour. Typically, labourers were hired on to learn the trade with a practising farmer, stayed a season or two, and then struck out on their own. Training in agricultural skills at the frontier was largely informal, and occurred in the private sector rather than in public educational institutions. The Canadian Pacific Railway, which recouped the cost of its transcontinental railway through land sale and monopolistic charges for freight to and from the farms, took some leadership in education by setting up ten model farms in Saskatchewan in 1884.

In the United States, the government accepted responsibility for agricultural education in 1862 when President Lincoln, the son of a dirt-poor farmer, signed the Morrill Act into law. This legislation gave thirty thousand acres of federal land to each senator and representative in Congress to sell in order to raise funds to create a public university for the teaching of agriculture and mechanics (engineering). The land carried a nominal book value of $1.25 per acre; this was supplemented in most states with local government and private money to start universities. In the home state of Justin Smith Morrill (1810–1898), the Morrill Act allowed the University of Vermont, founded as a privately funded institution in 1791, to obtain government funding. Iowa State (1862), Illinois (1867), and Texas Agricultural and Mechanical (1876) are examples of universities created under the Morrill scheme that retain strong programs in agriculture to this day. In 1887, each American state was authorized to create an Agricultural Experiment Station to work with the "land-grant colleges" to "conduct original researches or verify experiments on the physiology of plants and animals; the diseases to which they are severely subject, with the remedies for the same."[27] In 1890, land-grant colleges were created for black Americans in the southern states. In 1914, Congress passed the Smith–Lever Act, creating the US Cooperative Extension Service, which aimed to teach farmers directly through the land-grant college system.

In Canada, the mounted police took responsibility for providing a veterinary service for settlers in the early years. The Ontario Veterinary College, North America's first, was established as a private school in 1862 in Toronto (Fig. 23), and the Ontario Agricultural College was founded in Guelph in 1874 (Fig. 24). In 1922, the Ontario Veterinary College moved to the Guelph campus of the Ontario Agricultural College; thus North America's first veterinary college co-located with the first agricultural college in the Commonwealth outside of the United Kingdom. The OAC's founding agricultural curriculum was typical of the colleges of its time: students worked in the fields either for ten hours every second day in a six-day week, or each morning or afternoon as they alternated shifts between first- and second-year trainees. A typical annual cycle of practical experience would include tree-felling, stump-digging, plowing, seedling, hand-weeding, harvesting, and threshing, with the expectation that the academic side of the curriculum would be suspended during the busy weeks of seeding and harvesting.

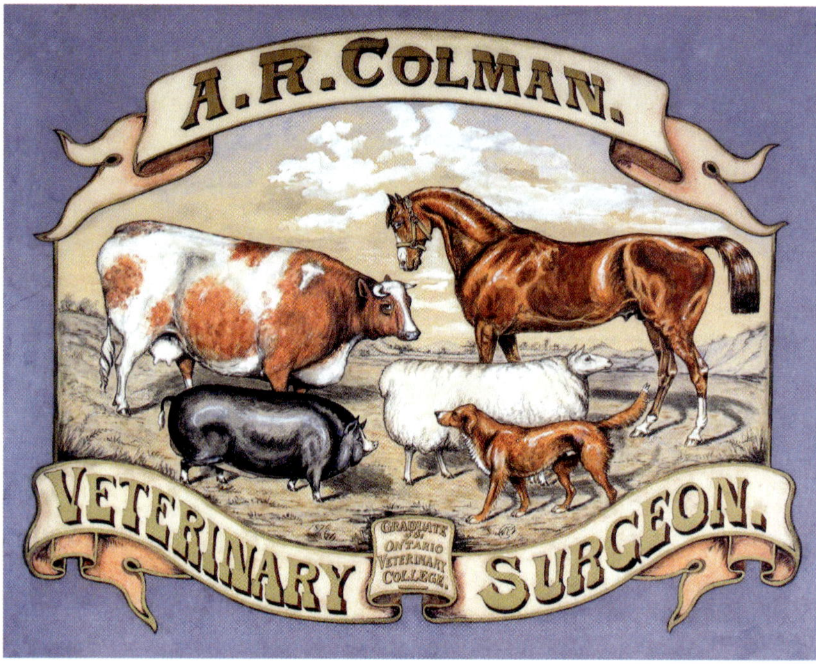

Figure 23
Alfred Russell Colman, *A.R. Colman Veterinary Surgeon,* 1876 (gouache and watercolour). Colman, an Ontario Veterinary College graduate, was so proud of his achievement that he designed and illustrated his own "diploma" in watercolour with gold leaf. Colman graduated with honours in 1876 and established his practice in the village of Jarvis, Ontario, where he continued to work until his death at age 85. His beautifully drawn painting served both to establish his credentials and as an advertisement to attract customers.

While these initiatives in formal education and extension for agriculture in the late 1800s might seem to have arrived late, there were few Old World precedents to guide them. Although learned societies had arisen in Europe in the early 1600s, the first national, state-funded agricultural extension service was announced, by France, only in 1879, seventeen years after Abraham Lincoln's initiative and only two years before the United States moved to create its federal extension service. Similarly, agricultural universities in the Old World did not predate their New World counterparts by many years: although a chair of Agriculture and Rural Economy was established at Edin-

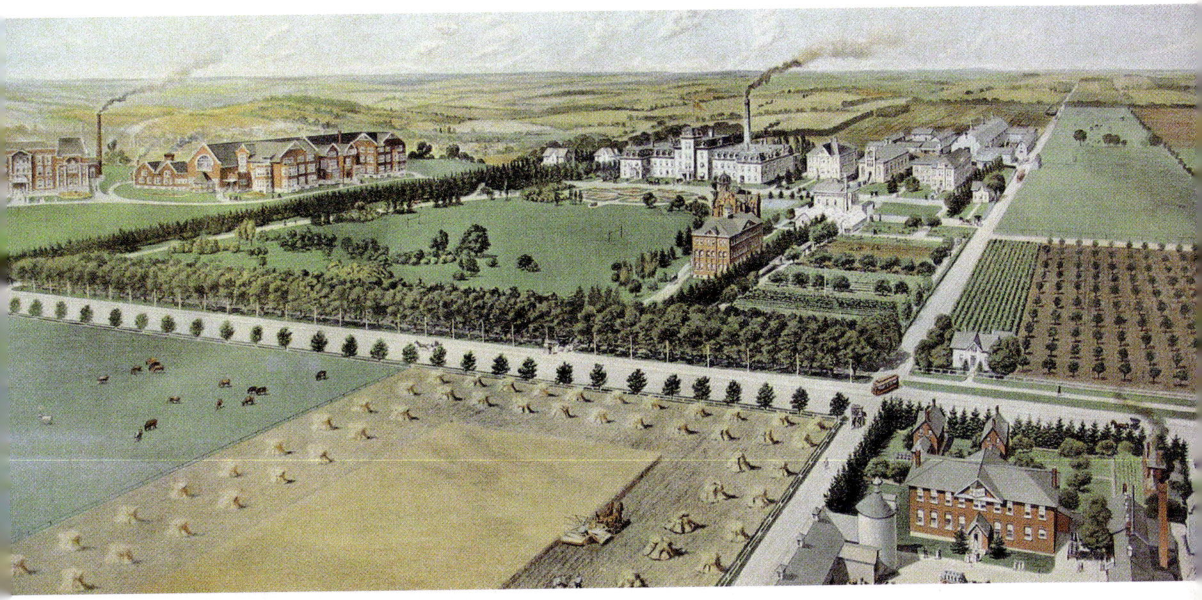

Figure 24
Charles Manly, *The Ontario Agricultural College, Guelph, Canada*, 1906 (watercolour). This bird's-eye view shows an orderly layout of buildings focused on a main administration building overlooking rose gardens and the college green. Experimental and teaching gardens surround the buildings.

burgh University in 1790 and at Oxford University in 1796, the first dedicated institution, the Royal Agricultural College at Cirencester, Gloucester, took its first twenty-five students only in 1846.

Schemes to import and train labour abounded, with little impact, from the 1800s to the early twentieth century. After World War I and a subsequent period of high unemployment in Europe, the (British) Empire Settlement Act of 1922 provided £3 million for each of fifteen years to approved training and settlement schemes in Canada. In the education sector, Macdonald College, formed in 1907 as part of Montreal's McGill University, imported thirty public school boys each year from 1924 until the Great Depression, as a contribution to providing an educated workforce for agriculture.

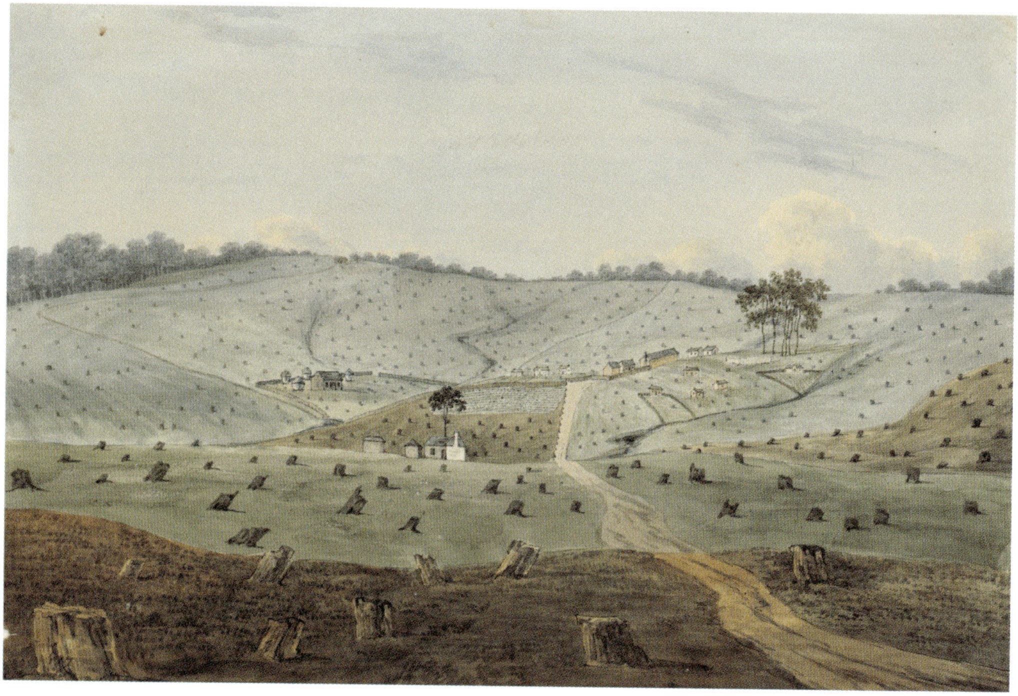

Figure 25
Government Agricultural Establishment, Castle Hill, circa 1806 (watercolour). An early painting of the governor's summer farm west of Sydney shows the ugly state of the property before "grubbing out" of the stumps of felled trees. Such a factual view was unusual and in direct contrast to the picturesque watercolours by amateur artists that presented Australia as idyllic Britain in disguise.

In North America agricultural settlement was first established on the coastal fringe, then followed rivers inland, and finally evolved into a broad expansion into the interior. In Australia, agricultural expansion followed a similar pattern. Sydney, settled in 1788, was geographically confined until about 1809, when the colonial administrator William Paterson established farms in the "forest lands" of the Hawkesbury Valley. Early paintings indicate the extent of crude, clear-felling of the forest (Fig. 25). Every tree was cut down by convict labour, leaving a landscape dotted with unsightly tree stumps that might, unless dug out and burned, take a half-century to rot away.

It might be said that the Australian convict experience was the world's first, and arguably the most radical, attempt at criminal rehabilitation. When convicts had served their time they were freed and, with good behaviour, would receive a grant of land so that they could set themselves up as smallholder farmers: something quite beyond their reach had they remained in Britain's class-conscious society. Progressively, freed convicts, poor immigrants, and failed farmers created a society of landless free labourers. They would sometimes live for generations on the same property, perhaps as shepherds on large "stations" or ranches. Others would shift for themselves as itinerate labourers or drifters. The vast majority had little or no formal education, although some were astonishingly creative. John Shaw Neilson (1872–1942), not atypical of his class and time, attended school for a total of about two years in the Wimmera district of Victoria. Nonetheless, throughout his life as an unskilled, drifting farm laborer, he wrote poetry:

> Quietly as rosebuds
> Talk to the thin air,
> Love came so lightly
> I knew not he was there.
>
> Quietly as lovers
> Creep at the middle moon,
> Softly as players tremble
> In the tears of a tune ...

And, drawn from the experience of seasons of hardship:

> Let your song be delicate.
> The flowers will hear:
> Too well they know the tremble
> Of the hollow year.[28]

However, as Neilson and others would learn to their cost, crops and farming techniques were often inappropriate; farming commonly consisted of continuous cropping of wheat or barley in winter and maize (if there was enough rainfall) in summer, all on soil low in nutrients. The rapid run-down in soil fertility, recurrent flooding of the river terraces that were favoured for cropping, and wildly fluctuating local prices caused fields to be abandoned, and new ones cleared, after three or four years. This was hard and wasteful work. It also constrained the size of cultivated holdings. The small scale of farming operations – together with erratic, declining yields – caused food shortages such that the colony periodically depended on wheat from Bengal and Rio de Janeiro.

Nutrient depletion, overgrazing, and intermittent droughts and floods are given as the stimuli[29] that prompted explorers to seek a passage over the Blue Mountains, which hitherto had hemmed in the colony. In 1813, amateur explorers Gregory Blaxland, William Wentworth, and William Lawson found a route from Sydney through the Blue Mountains to the rolling grassland to the west. Interestingly, this trio succeeded by beginner's logic in finding a passage to the interior where many others had failed: they hard-slogged it up heavily timbered ridges rather than following the valley floors, where earlier explorers had made quick progress only to be beaten by unscalable cliffs at the end of the valleys. Despite the proscription by Lachlan Macquarie, Governor of New South Wales, against settlement that extended beyond five designated "towns" in the interior, many sheep-farming families spread across the plains. By 1829 the colony of Sydney had expanded considerably; however, the city was constrained by government order to an area of 34,505 square miles:[30] a big expansion, but with lots yet to come across a country of three million square miles.

Once beyond the coastal fringe in Australia, as in North America, the vegetation gradually became less wooded, and travellers encountered extensive areas of grassy plain were interspersed with woodland and forest. In 1834, John Lhotsky (circa 1795–1866), an eccentric Polish naturalist, enthusiastically described the landscape near today's capital, Canberra:

> As far as the picturesque is concerned, nothing can be more beautiful. This great chain of plains ... are altogether destitute of trees of any kind, and only on the secondary hills and banks, which divide their placations, are some gum trees thinly scattered, whereas large timber covers the main ranges.
>
> O! that such a desert were my dwelling place.[31]

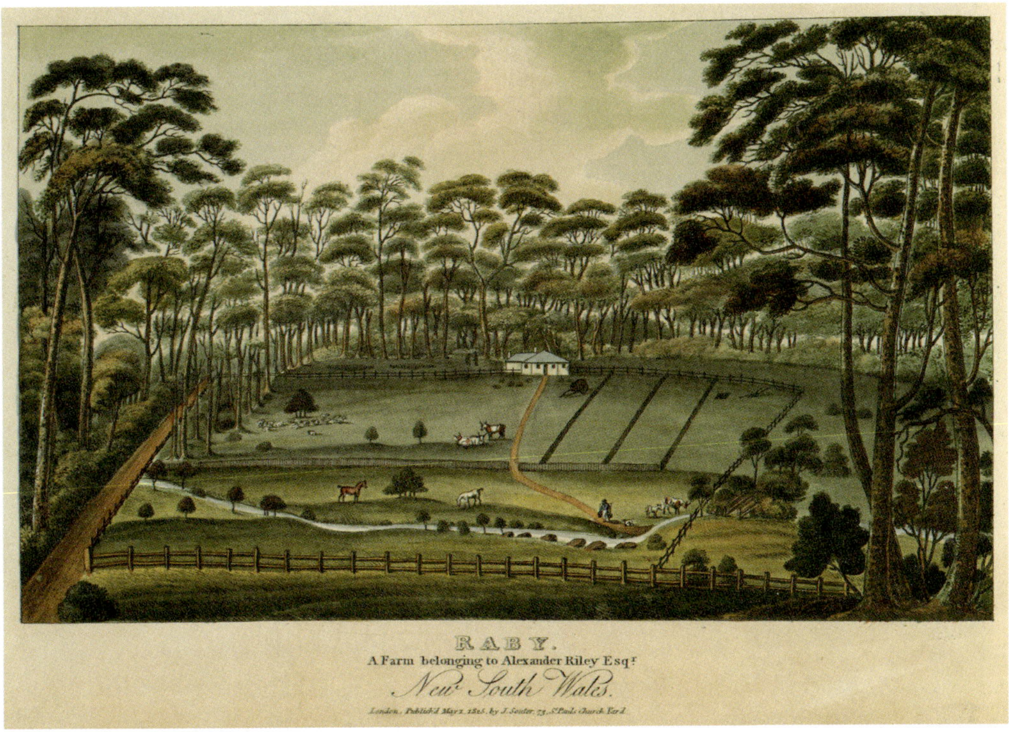

Figure 26
Joseph Lycett, *Raby, a farm belonging to Alexander Riley, Esq., New South Wales,* 1825 (etching and aquatint, printed in black ink from one copper plate, hand-coloured). Lycett, a former prisoner pardoned by Governor Lachlan Macquarie in view of his artistic ability, presents a romantic interpretation of this homestead, which seems like an oasis in the midst of a towering forest. A picturesque bridge mimics the landscaping features of the great houses of Britain.

In the early 1820s, settlement and clearing for cropping or livestock was carried out in areas controlled by the government. These areas were being surveyed and defined as the "limits of location." However, in 1834 the government lost control. In the period that has been called the "heroic" age of expansion, a flood of settlers travelled westward and squatted on unclaimed land. (A similar movement of settlers occurred in the United States in Virginia, Kentucky, and Ohio, and then further to the west.)

In Australia, the squatters took possession by marking areas, grazing cattle, and settling near surface water, which effectively controlled large areas of surrounding grazing land. Barbed wire, patented in 1874 by Joseph Glidden (1813–1906), an Illinois farmer, transformed the landscape by controlling the movement of livestock without the need for labour. There is an apocryphal story that Glidden tinkered with the wire to keep his dogs out of his wife's rose garden. During this era literally thousands of patents were being taken out on fencing and fence-related gadgets, all attempting to create effective, low-cost enclosures for large-scale farming. Weaver,[32] Christopher,[33] and Limerick[34] give more detailed accounts of the age of expansion in Australia, South Africa, and the United States respectively, than are possible here.

During the decades between 1870 and 1900, the rate of agricultural expansion levelled in the eastern and central United States (Appendix 1). However, in the west of the United States, Canada, and Australia the take-up of land for agricultural purposes began only in about the 1880s and continued until the period 1950–70. In Australia, the expansionist phase continued until the 1980s. One impetus for continued expansion was the formalization of land ownership that took place in a number of countries in the third quarter of the nineteenth century. Examples are the 1862 Homestead Act in the United States, the 1867 Land Act in Argentina, and the 1872 Homestead Act in Canada. This legislation was aimed at shifting government lands to private ownership by white settlers; in the United States and Argentina, it coincided with the government's determination to decimate native populations or consign them to marginal lands of little agricultural value. In 1878 the government of Argentina issued four thousand land bonds. Immigrant settlers were required to purchase a minimum of four bonds, entitling them to ownership of twenty-five thousand acres. The money was raised, explicitly, to support an army to exterminate the aboriginal peoples and thereby make more land available for settlement. General Julio Roca (1843–1914) achieved this in 1879, during his first term as President, in one of history's extreme examples of using land-sale to shift ownership through dispossession.

In addition to introducing agriculture into regions occupied by indigenous people who farmed only small areas, Europeans colonized regions, such as Madagascar, where there was no resident population. Elsewhere, however, the story of European settlement was one of the displacement of indigenous people: in parts of Africa, Indonesia, Malaysia, and the Philippines, settled farmers were displaced by European immigrants who introduced plantation agriculture with the aid of policies developed by governments strongly influenced by their colonists. A deliberate change occurred from village-based food cropping to owner-based plantations (particularly sugar and rubber). Plantations provided a cheap and controlled supply of raw materials for Europe's ongoing industrialization. This rapid change had a devastating effect on the food supply for indigenous peoples who did not operate in a cash economy and on the status and opportunities of smallholders, who now found themselves landless.

The age of expansion continued into the twentieth century at a rapid rate. In the Western New World, including Argentina, expansion proceeded into drier and more temperate lands, where wheat and beef were the main products. All manner of inventions helped to mechanize agriculture and make it less reliant on a supply of manual labour. Innovations such as cheap fertilizer, threshing machines (Fig. 27), and refrigeration enabled the transport of perishable goods to European markets, underpinned the late phase of expansion. A type of windmill patented in Australia in 1864 allowed for semi-automatic watering of stock, expansion of grazing, and limited irrigation of crops. The introduction of barbed wire made low-cost fencing possible for large-scale farming. The value of imports of annealed wire to New South Wales rose from £81,000 in the decade of the 1860s to £813,000 in the 1870s, thereby reducing the need for shepherds as well as allowing increasing stocking rates and flexibility of stocking.

Expansion in the drier west was not without difficulties, some of which persist today.[35] In southern Australia a succession of high-rainfall seasons in the early 1870s provoked agitation for agricultural settlement to be allowed further inland, to the northeast of the colony. The surveyor-general George Goyder (1826–1898) recommended against it, instead advising that land release and settlement based on cropping should be confined to areas receiving at least 250 millimetres of rainfall each year – a territory demarcated by "Goyder's Line." This advice was unheeded. With government support,

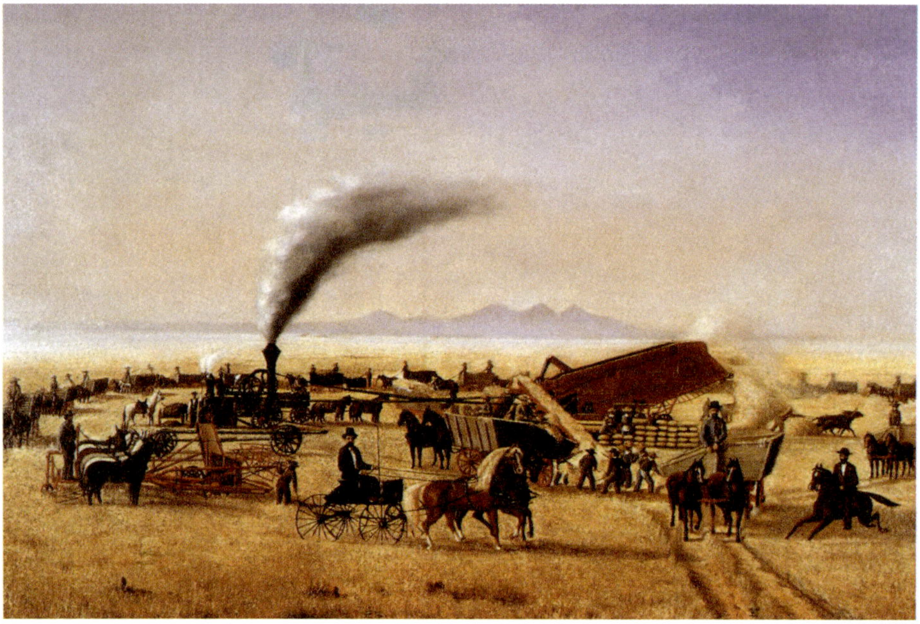

Figure 27
After Andrew Putnam Hill, *George Hoag's Steam Threshing Outfit and Crew Setting a New One-day World's Record*, 1878 (lithograph).

settlers took up land and new towns boomed. The hypothesis that "rain would follow the plough" gained serious, widespread acceptance. Today, the town of Farina is a ruin: the ochre-red remains of solid Victorian buildings stand roofless and crumbling on a flat stony plain dotted with a few peppermint trees. Marree, one of the few settlements on the arid side beyond Goyder's Line, survives as a transport centre. It is a mix of abandoned earth, timber, and galvanized iron shanties, a sleeping, dry ruin of a railway station, and a few houses and gardens struggling in the glare of the sun, occupied by aboriginal and whites who are casual workers on the surrounding ranches. Even at Marree, the farmers of the age of "rain following the plough" are a hundred years gone (Fig. 28).

As land was being opened up for settlement in Western Australia, the scientists of the day recognized that, on an old continent that once lay beneath the sea, there was the possibility that soil salinity would limit agriculture. In 1917, Professor William Patterson of

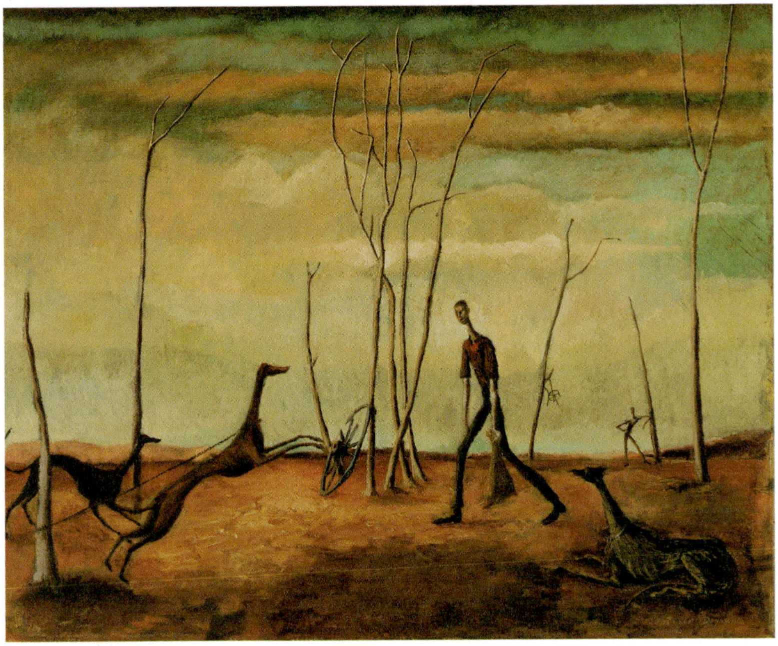

Figure 28
Russell Drysdale, *A Man Feeding his Dogs*, 1941 (oil on canvas). This glum painting is a sharp contrast to Joseph Lycett's romanticized view of an Australian farm (Fig. 26). Drysdale depicted the devastating effect of drought on agricultural communities in the 1930s and 1940s, helping to bring the issue to the attention of a broad public.

the University of Western Australia warned an enquiry against large-scale expansion, only to be rebuffed by the report of the Royal Commission on the Mallee and Esperance Lands: "The Commission having given the question close consideration strongly urges that scientific prejudice against our mallee lands be not permitted to stand in the way of their being opened up for agricultural purposes."[36] It was not until 1981 that clearing restrictions were imposed in southwestern catchments. Now that the era of expansion has ended and the area of Copeland recedes and salty lakes increase, Patterson's prophesy returns to us, along with assessments of an almost intractable problem of dying land and townships.

Whereas the American, Argentinean, Australian, and Canadian expansions were driven by immigration from Europe, Brazil is a different case – and a topical one, for it represents the last large area where the paradigm of agricultural expansion is still alive. Although the Amazon was colonized by Portuguese, French, Dutch, and British about 1615, its changing landscape is the outcome of an expansionist movement based on local population growth and internal migration.

As in North America, the settlement in South America from the seventeenth to the nineteenth centuries proceeded along river banks. The only immigrant town of substantial size was Belém, located at the mouth of the Amazon. Belém's population of 33,000 in 1850 grew to 275,000 by 1912. Colonists established sugar plantations in the northeast and farmed the coastal region. The interior basin was dense forest and was considered a site of unknown disease, giving little motivation for settlement until the mid-1800s. There were very few settlers in the interior; trade was complicated and river-based, and the only viable exports were products from turtles farmed in the villages. However, from 1860 to 1910 there was local wealth without cultivation in the conventional sense, because the Amazon supplied most of the world's rubber.

Then, in 1888, slavery was abolished in Brazil. Plantation owners had difficulty finding cheap labour. The price of rubber dropped after 1908 as Malaysia entered the market with plantation-grown rubber. By 1911 the rubber market was oversupplied, and the Amazon Basin was in an economic depression. Government attempts to develop infrastructure and agriculture were ineffective. A railway link to Belém helped the city's economy but did little for the countryside. As a result, the few settlements clung to the riverbanks – in stark contrast to North America and Australia, where railways opened up the interior. In the lower Amazon, immigrants established and managed cocoa plantations before and during the rubber boom.

In 1970, Brazil established the National Integration Program, which was intended to foster infrastructure development that would lead to agricultural settlement in Brazil's northeast and the Amazon Basin. The annual rate of forest clearing has been dramatic, accelerating from 11,000 square kilometres in 1990 to 18,000 in 2000. However, as the Brazilian government has indicated, these recent clearings need to be put in the context of the longer term: 600,000 square kilometres have been cleared since colonial times. Assuming a further 340,000 kilometres are cleared in the next twenty years, this will amount to, overall, about one-quarter of the Amazonian rainforest. Today, while the central Amazon and its clearing understandably draw world attention, it is the Atlantic forests – now only a thin strip along the entire coast, where much less forest habitat remains – that are of even more concern from the viewpoint of landscape and biodiversity.

In the age of agricultural expansion settlers brought with them their potted plants, seeds, birds, and livestock. They planted and delighted in what they recreated, thousands of miles, and months or years, from "home" (Fig. 29). Attitudes toward the local environment and fauna were, as today, multi-faceted. On the one hand, there was an enthusiasm for hunting and fishing – a frontier mentality – both as sport and as a means of obtaining food. Most people, whether immigrant or native, were much more concerned with securing a supply of food than with protecting the fish and animal populations that sustained it, even if this meant over-fishing or over-hunting. However, colonists were quick to protect native flora and fauna when it suited them. Australia's first environmental laws were enacted in April 1788 to protect plantains just four days after their discovery.[37] This measure was motivated by a keen interest in preserving a supply of fresh food, rather than by any twenty-first-century concept of biodiversity. Birds, on the other hand, were such good sport that attempts to limit their killing to the number that might reasonably be needed for food almost never succeeded; colonists harvested hundreds of thousands in a season, causing the extinction of species long before their habitats became endangered. In North America, bison were not protected until after they were effectively extinguished by 1880.

Not all treatment of the land was rape and pillage, although these might be appropriate verbs to describe the Europeans' treatment of the indigenous peoples of North American and Australia. In terms of the landscape, settlers cleared land indiscriminately causing the eradication of some animal species. However, there were glimmers of enlightenment, as exemplified by a report from Lower Canada (Quebec) in 1849: "The right of property is certainly sacred and inviolate: but the soil only belongs to man on condition that he works

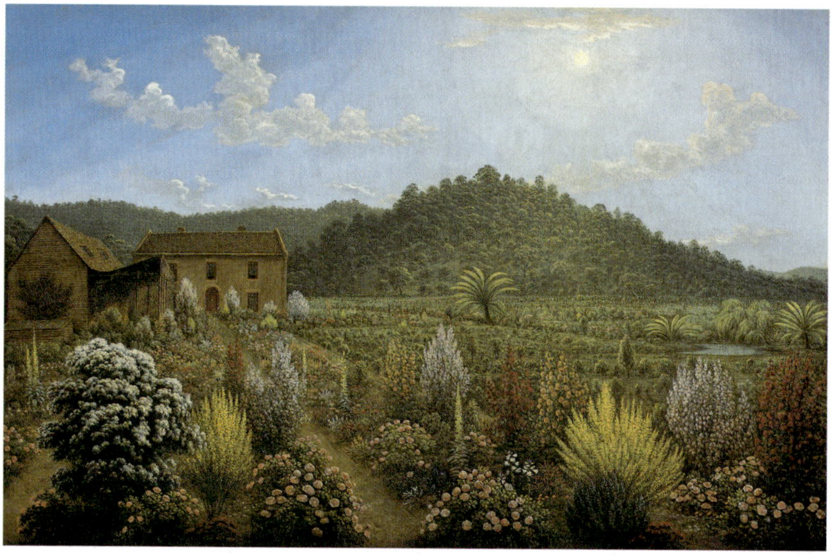

Figure 29
John Glover, *A view of the artist's house and garden, in Mills Plain, Van Diemen's Land, Deddington, Tasmania,* 1835 (oil on canvas). Glover was a self-taught artist who emigrated in 1830, at age sixty-three, from Leicestershire, England, to settle his family on a farm on the Nile River near Launceston, Tasmania. This portrait of his farm shows his pride and sense of order. The plants in neat garden plots are importations; their purpose, to bring an English sensibility to a foreign land and to create a local supply of food. Although the foreground of the painting is dominated by flowers, we might reflect on the orderliness (or indeed, extraordinariness) of the gardens, featuring a variety of fruits and vegetables.

and cultivates it; and possession carries with it the obligation to make use of what one possesses in such a manner as to not injure others. Property should have its duties and obligations as well as its rights."[38] Here, those duties were related mostly to tilling the land and feeding the colony: this was an early policy to discourage out-migration from Quebec to the United States. Nonetheless, this counsel to "not injure others" predated the ideology of land conservation and sustainability by 130 years. In 1926, American historian Avery Craven (1885–1980) published a thesis on the soil exhaustion that followed the settlement of New England.[39] It is a sad irony that Craven's book on soil degradation appeared just two years before the double catastrophe of the Great Depression and the dustbowl years of the southwestern United States. Ranchers in the high plains of

Colorado, Texas, Oklahoma, New Mexico, and Kansas had stocked cattle excessively, causing overgrazing of the land. Overstocking was probably a part of their usual agricultural practice, but was no doubt worsened by the financial depression, which resulted in fewer animals being sold.

Similarly, single-furrow ploughs were used – even when the soil degradation was occurring – to turn natural, relatively low-rainfall grassland into wheat fields. The fields were uncovered at the beginning and end of each growing season. In 1931 there was a bumper crop. Then, in 1932–33, below-average rainfall caused severe loss of grass and wind erosion. Dust-storms – up to hundreds each year from 1933 to 1937 – darkened the region. Many tales of hardship and survival arose from this era, ranging from the misadventures of travellers in dust storms so severe that they could not see their own feet to crop failures and civic disaster. Woody Guthrie's "The Great Dust Storm" expresses the psychological impact of these natural events:

> It fell across our city like a curtain
> of black rolled down,
> We thought it was our judgement, we thought
> it was our doom."[40]

In his history of American agriculture, R. Douglas Hurt gives a sense of the scale of this era's catastophes: "In May 1934 a dust storm removed an estimated 300 million tons of soil from the Great Plains" and ... "Perhaps 500,000 people, mostly farmers, pushed by the worst years of the drought, left the Great Plains during the 1930s"[41] (Fig. 30).

Some took hard lessons from environmental disaster. Two men who were growing up on farms during the prairie dustbowl years and in the era of rust-ravaged crops in Australia were Charles Quinby and Irvine Watson. Quinby worked for the United States Department of Agriculture in Texas; Watson came from the central west of New South Wales, in gently rolling wheat and sheep country. In old age Watson recounted his family's sense of desolation – silence at the dinner table – in the year when initially the fields were filled with

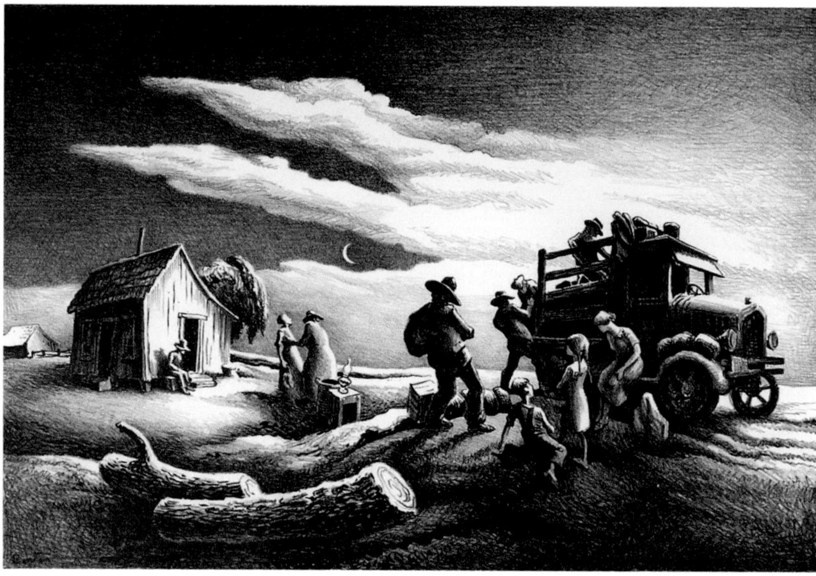

Figure 30
Thomas Hart Benton, *Departure of the Joads,* 1939 (lithograph). A family packs their few belongings to leave a hopeless farmstead in an attempt to find a better life.

good-looking wheat crops that, as the season progressed, finished covered with reddish pustules and without grain. Such experience committed Quinby and Watson to agricultural science: as the years passed, these two soft-spoken former country boys developed, respectively, the hybrid dwarf sorghum on which all modern sorghum production is based, and world-leading knowledge of rust and rust-resistant wheats.

Agricultural landscapes were also altered by the direct human intervention of the clear-felling of trees, which had been a normal practice in Europe up to the eighteenth century. In the New England states, it involved ring-barking each tree, cutting it down with an axe, and burning the stump and the residue after anything useful had been retrieved for building, fence-posts, or firewood. The practice became more problematic as mechanization increased the capacity to deforest large areas. Today in Australia and Brazil a hundred hectares can be clear-cut in a day by means of tractors operated in pairs, tied together with heavy-gauge chain, but in the late eighteenth and early nineteenth centuries, clear-felling on sandy, windswept soils throughout the New Worlds had already created exposed landscapes that were filled only seasonally with crops, leading to problems with wind and water erosion. In parts of southern Ontario, forested landscapes gave way to drifting sand. In many areas, clearing by cutting and burning turned richness into wasteland. In our own era, the agricultural science community has been faced with the legacy of land clearance: "Southern Ontario has been cleared and tilled for less than one hundred and fifty years ... and yet we are confronted with many serious problems of soil and water, owing to the lack of forest cover."[42] But those effects were foreseen by some agriculturalists even a century ago. In 1879, the Ontario Fruit Growers' Association "carefully instruct[ed] the farming community [on] how much depends on the judicious planting of forest trees, their presence producing abundant rainfall, preserving and distributing moisture and thereby forming a preventative against drought and devastating floods."[43] Gradually, the problem was acknowledged by the public and by governments, resulting in the designation of some areas as federal nature reserves and provincial parks where the public could enjoy nature "in the wild" – such as at Yellowstone National Park in the northwestern United States (1872) and Algonquin Park in Ontario (1893). The public also took the initiative to create privately supported woodlots and regeneration areas. Thanks to the solidarity among scientists, amateur naturalists, and artists, clear-felling in national parks was discontinued, and agricultural and park landscapes have changed.

One example of this restorative trend is Brown's Wood, on the University of Guelph campus in a suburban town within commuting distance of Toronto. This five-hectare forest is surrounded by houses, college buildings, and a parking lot. It exemplifies the outcome of the contemporary push for conservation and regeneration. Planted by William Brown, the first professor of agriculture at the Ontario Agricultural College, Guelph, and his students in 1887, the Wood is now a mature, second-generation, forest. It changes its dress with the seasons: sparse in winter, a layer of snow accentuating the mounds of parallel hillocks on which it stands. The hillocks do not date to glacial times, but to wind-blown sands of the late 1800s. The adjacent field, fuelled by another hundred years of technical improvements associated with productivity, annually produced a crop of wheat or corn. The field, only ten hectares in size, looked much like the flat field of southwest Australia mentioned at the beginning of this chapter. Last year tractors shifted the sand about, and apartments were built. Corn production in suburbia nowadays has little value.

Depression and dustbowl suggest a simple answer: government-supported construction of dams for irrigated agriculture (Fig. 31). This is the epitome of the age of expansion, when humans can shape the landscape to their needs. The Boulder Dam (renamed the Hoover Dam in 1947) on the Colorado River was completed in 1935 and filled by about 1938. Its mission is largely flood control and the generation of hydroelectricity. It is not clear how much of California's irrigated agriculture, which includes most of that state's top twenty industries, such as dairy, grapes, cotton, and fruit, depend on water from the Boulder Dam.

The Aswan High Dam, filled in 1970, has been "the cornerstone for Egypt's sustainable agricultural plans,"[44] protecting against severe flooding in the late 1970s and drought in the 1980s. The dam allows irrigation of an extra five million acres of cropland. Water loss from its surface amounts to between 7 and 12 km^2. Sedimentation, which was feared to be a threat to its long-term viability, seems to have been over-estimated, with estimates that it will take five hundred years to fill the "dead storage" of the lake. Salinization of the agricultural landscape remains an ongoing concern, as in most semi-arid regions.

Although vexing social problems have been created by the construction of the Aswan dam, particularly the resettlement of a hundred thousand Nubians thirty-five kilometres north or east, and the submerging of archeological monuments, it seems clear that it fulfills its intended function, namely to increase the viability of Egyptian agriculture and its food support for a rapidly expanding population. By contrast, the Kariba Dam, impounding Africa's largest artificial lake on the Zambesi, filled in 1963, is mainly for hydroelectricity for

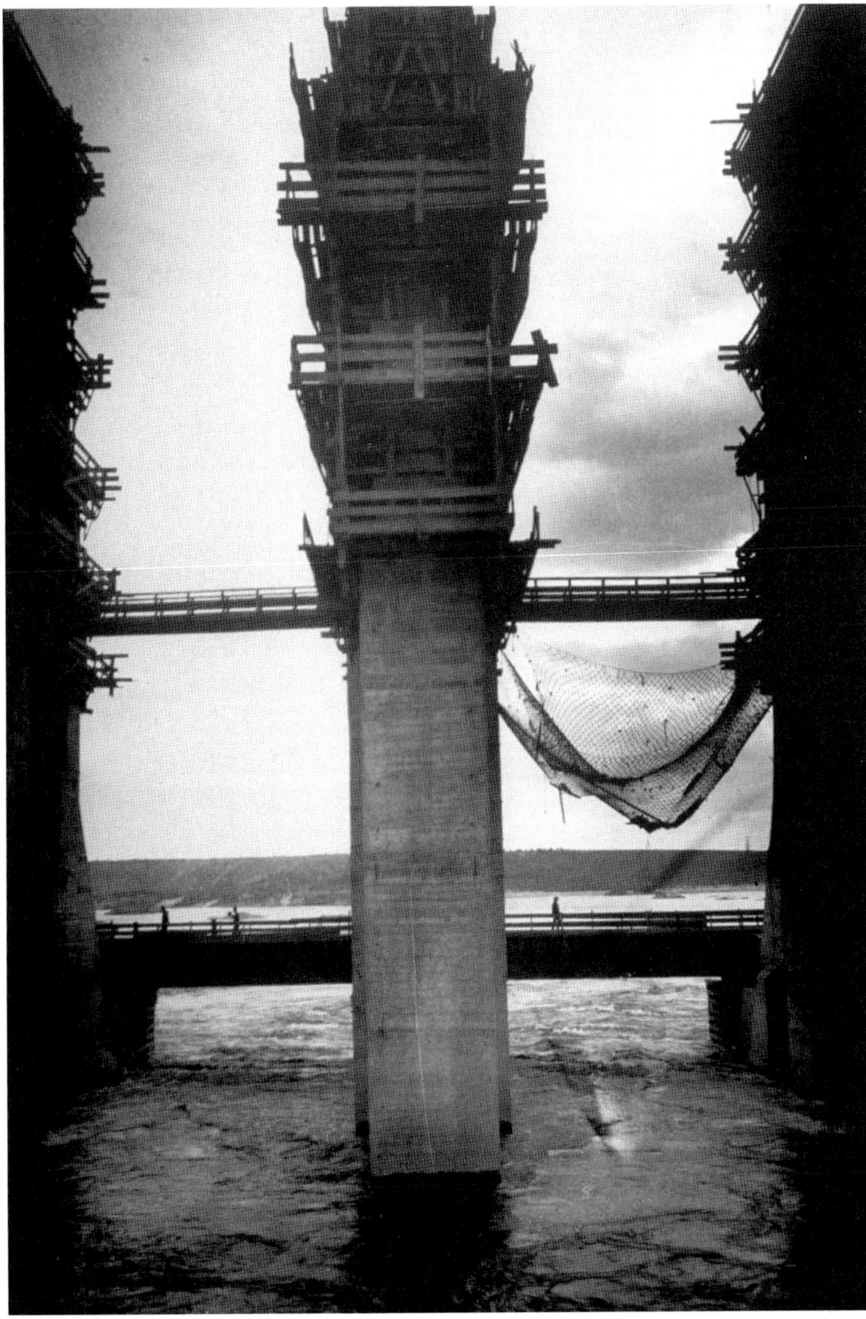

Figure 31
Margaret Bourke-White, *World's largest dam across the Dnieper River under construction*, 1931 (silver gelatin print). Photojournalist Margaret Bourke-White's photograph emphasizes the heroic verticality of the piers in the dam under construction in Russia. The 1930s ushered in the streamlined modern style that emphasized efficiency, science and progress, and a view that nature could be conquered by human ingenuity.

nearby mining industries and tourism. It is surprising, flying over Kariba, to see so much water and so little agriculture.

The era of expansion may be simplistically summarized into overlapping sequences of discovery and hunting, fishing, trading, ranching, and crop-based agriculture. The first phase coincided with some adoption of the aboriginal way of life: the fur-traders and their trade routes and networks in Canada are so similar to the ways of the nomadic Indians that it is not surprising that, in Canada more than anywhere else, there was a close interchange, including joint military ventures, between European traders and settlers and Indians.

The second phase, of ranching, was in some regions an end in itself: a viable industry that is sustained today. Elsewhere it is seen as a transition, within which society was based on family units, usually living in isolation and without supporting infrastructure such as roads, railways, health services, or schools. The later phase of cultivation resulted in vast transformation of the landscape, and the greatest change in public perception of what was a "good" landscape. This final phase reached its peak in irrigated agriculture and the construction of the world's large-scale dams and irrigation projects, as in Boulder, Kariba, Aswan, and today in the Yangtze.

3

PRODUCTION AND PRODUCTIVITY

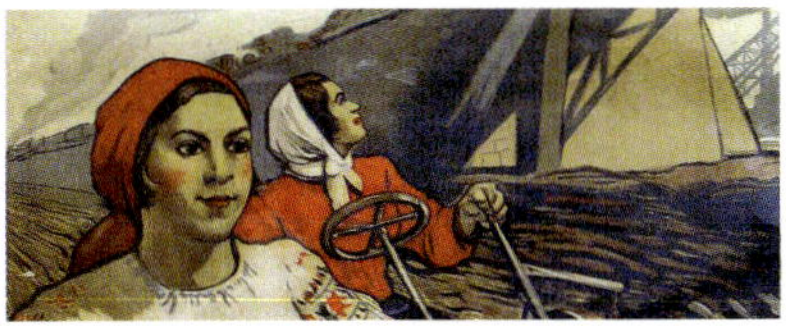

From the mid-1800s onward, the agricultural paradigms of production and productivity began to overlap. In Europe, the widespread use of enclosures together with changed inputs to agriculture (particularly the payment of labour) created conditions in which innovation could flourish. Similarly, in the western New Worlds, land settlement and the rapid expansion of agriculture encouraged entrepreneurship and innovation. Severe winters (in the north) or summers (in the south) together with naive and inappropriate farming techniques were problematic. For instance, farming practices suitable to Europe, which has a fairly high rainfall, were applied in the more seasonal and drought-prone climates of Australia, South Africa, and Argentina. This meant that at the turn of the century food security was still not to be taken for granted. The population of Europe had grown substantially, and an increasing percentage of the population was moving to towns. Intermittent serious food shortages occurred both in Europe and in the New Worlds. The landless poor struggled (Figs. 32, 33). Both the increase in population and the need for a guaranteed food supply spurred innovation for personal profit.

Wars created additional pressures by hampering trade and causing substantial increases in the price of grain. Millet's *The Gleaners*

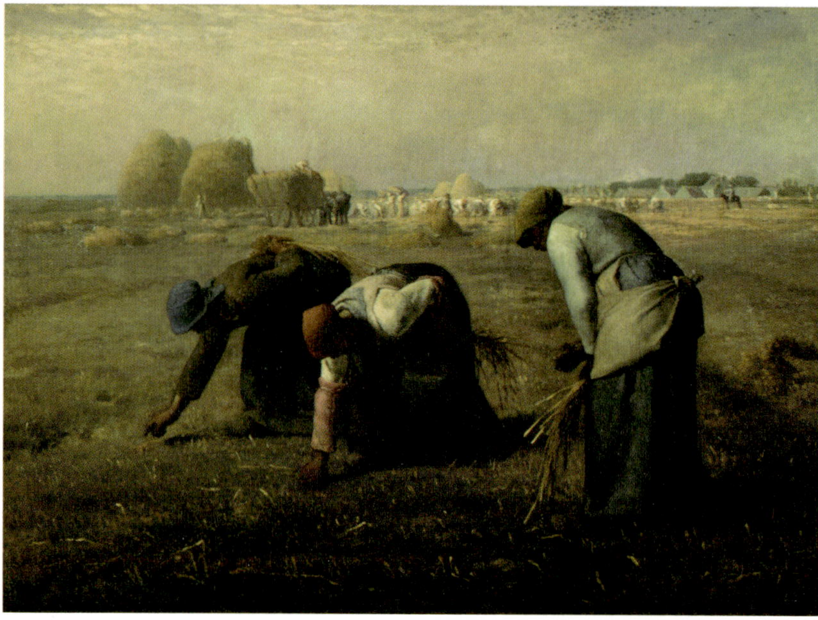

Figure 32
Jean-François Millet, *The Gleaners,* 1857 (oil on canvas). Millet draws attention to the poorest of the landless people of Europe, gleaners collecting fallen grain after harvest. The apparent harmony of the women gleaning with the sunlit grain and hay in the background is deceptive: in publicizing the scrounge-for-existence lifestyle of the poor, this painting prompted a critic to suggest that it carried "the thorn of revolution and the guillotine of 1793."[1]

Figure 33
Käthe Kollwitz, *Pflugzieher und Weib (Klipstein 61)* [*Plowmen and Woman*], 1902 (lithograph). Kollwitz offers a grim view of the devastation caused by war: poverty, starvation, displacement of survivors, and the collapse of the agricultural systems.

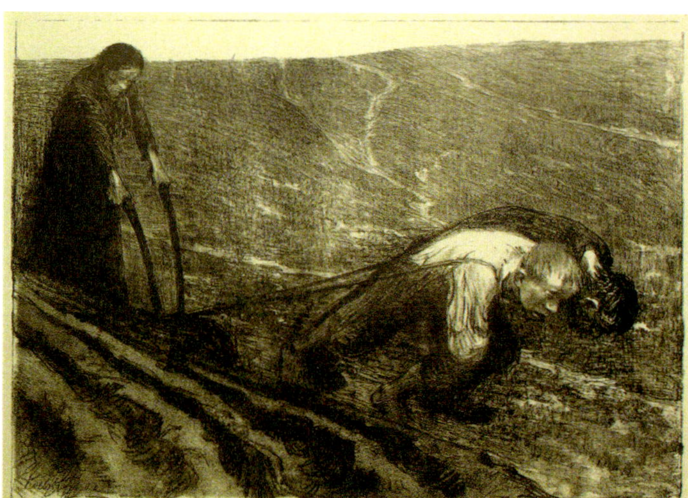

epitomizes the 40 years of trouble in European agriculture, following the Napoleonic Wars. When the wars ended in 1815, there were significant harvests and imports to Britain, where the price of wheat and other grains more than halved between January 1813 and December 1815. Farmers emerging from the manorial system were largely smallholder tenants and had neither the cash nor the technology they needed to diversify. And so they reacted perversely, as farmers still do today: as prices fell, they planted more crops. They shifted to growing wheat, which was seen as more profitable than other grains, and ploughed up grazing land. As one landowner informed the 1836 Agricultural Committee of the House of Lords, farmers had "scourged their land more and cropped it harder and broken up more lea [fallow or grassland] than they ought to have done."[2]

The land area sown to cereals increased, and the yields per hectare rose, factors that may have contributed equally to overproduction and continuing low prices. Prices for grain were predominantly poor in England in 1821–23 (after a particularly large expansion in area under cereal cultivation in 1819 and 1820) and again in 1835 (following a bumper harvest that year). The very low prices of the mid-1830s finally created a climate in which cereal production was not economical for some farmers: the more they planted, the more money they lost. Ultimately, low prices had four outcomes: (1) increased poverty among poor farmers and landless people, along with their migration from agricultural to urban livelihoods; (2) a decisive reduction in the total area devoted to cropping as less efficient farmers or those in marginal locations reverted to other land uses; (3) regional specialization in agriculture from cereals to other crops or to livestock; and (4) a search for greater efficiency and innovation. Thus, the mid-nineteenth century was the time to give attention to how much food could be produced per unit of land and to the efficiency of that production: how much food at what input cost – in a word, "productivity."

The plight of poor farmers who were ill-equipped with respect to skill or social standing to integrate successfully into urban environments was depicted by many artists. Lithographs such as Théodore Géricault's *Pity the Sorrows of a Poor Old Man* (Fig. 34) were inexpensive and widely distributed, and had the effect of arousing public sympathy, leading to the establishment of at least a few charitable organizations. Artists such as William Hogarth (1697–1764), James Gillray (1756–1815), and Thomas Rowlandson (1756–1827) produced prodigious numbers of caricatures and satirical illustrations that lampooned pomposity, excess, and cruelty.

Intermittent shortages of food alternated with gluts that bankrupted farmers. An interest in food production was heightened among city folk as its consumption became more precarious. In 1800, most of the family income of urban population was spent on food: 60 per cent is a commonly estimated figure, in contrast to 2000 when food purchases consumed only about 10 per cent of family incomes in North America. Given the limited understanding of hygiene and food safety, food-borne illness and death were commonplace. In nineteenth-century London,

> lice and even bugs infested the houses of rich and poor alike. Pests of myriad flies were bred by the festering garbage of the streets. Food was contaminated to a monstrous degree. Milk was sold in open containers, and had usually been diluted with the water from the iron cow, the tap in every diary. Bread was poisoned with alum, chalk and other whiteners. The constituents of beer were more than suspect; but of course all this pressed more heavily upon the poor in their crowded and filthy ghettoes than on the rich in their fine houses in the West End.[3]

One of the first to make a major scientific contribution to the understanding of food safety was Louis Pasteur (1828–1895). Pasteur made one of the most important findings in medical history when he discovered that most infectious diseases are caused by germs. Pasteur discovered that by applying heat to perishable food (the process now called "pasteurization") one could kill harmful microbes without harming the food. Another contribution to food science was his understanding of the process of fermentation, an important component of wine-making and brewing. As important as it was to attain a rational understanding of such commonplace things, the growing vogue for scientific investigation did not escape the jabs of the satirists (Fig. 35).

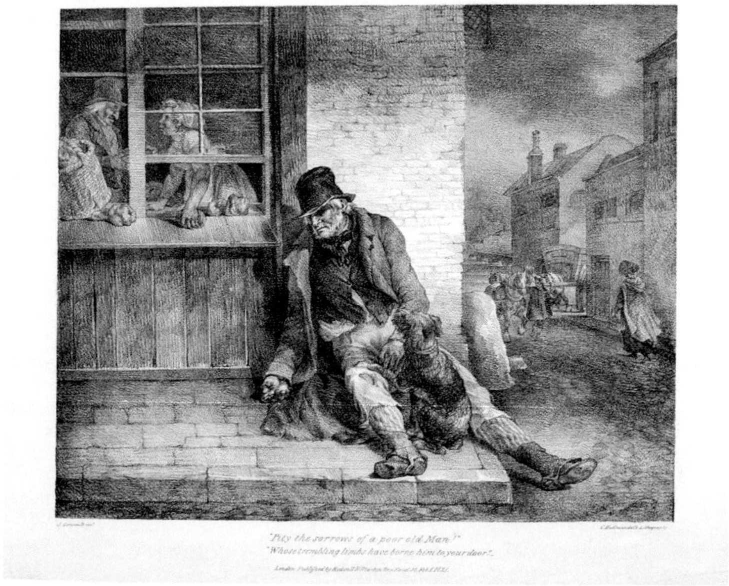

Figure 34
Théodore Géricault, *Pity the Sorrow of a Poor Old Man,* 1821 (lithograph). A beggar is ignored by passersby while in the background a farm cart delivers produce to market.

It is fortunate, however, that in mid-nineteenth century Europe there was a growing interest among the commercial and academic elite in the serious problems facing the farming sector: low grain prices, poverty and food insecurity, the improper ploughing-up of grasslands, the extension of agriculture into unsustainable locations, and the increasing cost of farm labour. Had these factors not also coincided with a new age of scientific discovery in biology, chemistry, and other disciplines that we recognize today as underpinning agriculture, they could have precipitated even worse hardship. But, while the need to increase farm productivity stimulated an age of scientific agriculture that has gathered momentum to the present time, we should acknowledge that careful study and writing on "good agriculture" – such as the Fitzherberts' *Booke of Husbandrye* of 1523 – date from the sixteenth century.[4] A century later the Royal Society was founded in London in 1660; the landed gentry and

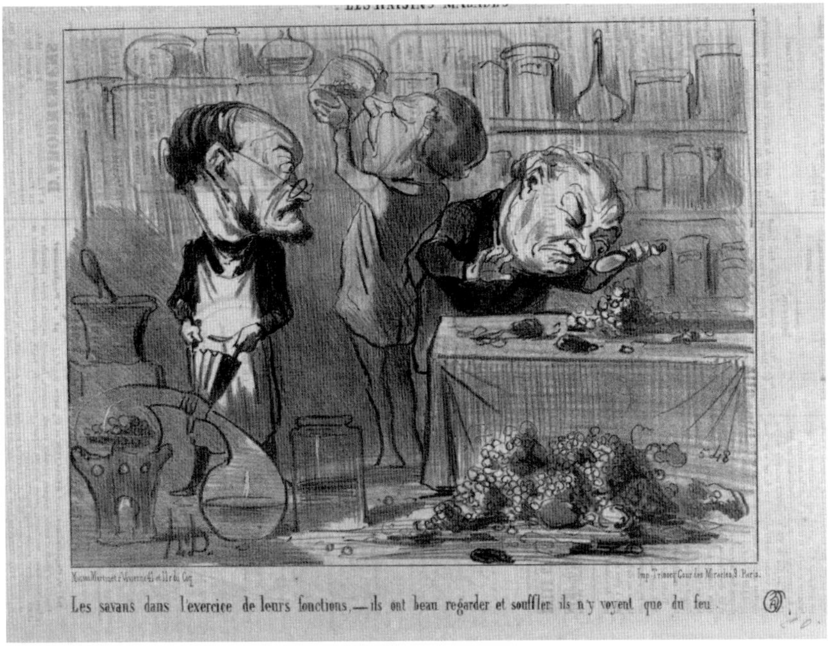

Figure 35
Honoré Daumier, *Les savants dans l'exercise de leurs fonctions* [*Skilled gentlemen in an exercise of their functions*], 1853 (lithograph). Daumier satirizes the pretensions of the *petite bourgeoisie* in nineteenth-century France, including these would-be botanists examining grapes with a magnifying glass.

lords among its members ensured the society's financial survival as well as stimulating the desire for scientific enquiry. The Royal Society for the Arts was formed in 1754. Among its six standing committees, two were dedicated to the betterment of agriculture and technology and offered cash prizes for innovation in agriculture.

The English agricultural reformer Thomas Coke (1752–1842) played a significant role in the promulgation of information and inventions. In an era full of bright ideas, some wacky and others downright dangerous (such as using early steam engines in tillage operations), there was a need to demonstrate and disseminate more practical innovations. Coke inherited a farm in 1776 and straightaway began annual "sheep shearings," the forerunner of today's extension or field days. Following Coke's widely-known example, the

English (later, Royal) Agricultural Society held its first "show" on July 1839, where landed gentry sporting top hats and canes presented their best and fattest livestock and newfangled equipment. There were classes for Shorthorn, Devon and Hereford breeds of cattle, and for "any others." Classes for pigs, sheep and horses did not become prominent until the 1860s. It took until then for dairy breeds to be delineated; Jerseys and Guernseys were recognized as separate breeds in 1871, and Friesians in 1911. It is interesting to reflect on how mainstream identities in agriculture, such as the Friesian or Holstein cattle on which our milk supply is based, went unrecognized as distinct breeds less than a century ago.

The goal of agricultural fairs was to encourage the spread of farming knowledge and to showcase new technology. The curiosity of the many townsfolk who were attracted to the early shows in Europe and the New Worlds ensured their viability; about twenty thousand visitors are thought to have attended the first English show in 1839, and their numbers rose to over a hundred thousand by the 1870s. With the passage of time, however, the increasing sophistication of urban populations together with the growing physical and social distance between city dwellers and farmers led to a change in the nature of farm shows, which began to fall into tough financial times.

Agricultural fairs exemplify the paradigm of production: they are intended to showcase an outstanding quantity and quality of animals (alive or dead!), grains, and vegetables, and the yield per animal or hectare, regardless of cost. Indeed, early English shows were criticized for displaying animals so fat they were unable to walk. A *Punch* cartoon of 1865 showed a prize pig, fattened until it was spherical, with poles attached to its legs so it could walk, rather than roll. Another example is Thomas Weaver's depiction of an enormous short-horned heifer in a cattle shed with classical columns (Fig. 36). This preoccupation with size continues today. At the Royal Winter Fair in Toronto, prize-winning squashes are a reliable crowd-pleaser: huge, bloated, knobbly pumpkins each weighing about 500 kilograms are carried by forklift to a roped-off carpeted podium, where they continue to demonstrate what agriculture *can* produce, not necessarily what it is efficient to produce, nor what consumers want. Many agricultural shows have progressively transformed

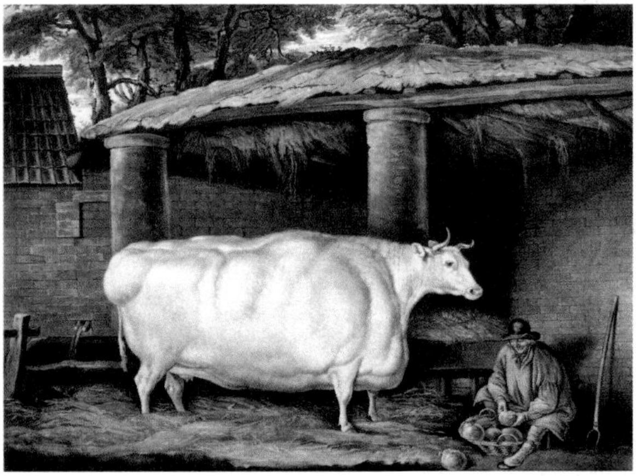

Figure 36
Thomas Weaver, *A Short Horned Heifer, Seven Years Old*, circa 1811–19 (engraving). It was common in this period for landed gentry to commission engravings of their houses and prized livestock.

themselves from showcases of innovation to entertainments with agricultural themes (Figs. 37, 38).

Productivity is the ratio of "output" (e.g., tonnes of wheat) to "input" (e.g., the cost of land, fertilizers, labour, and machinery). This ratio is used to assess the efficiency and sustainability of agricultural systems for the purposes of household economy, tax collecting, or energy use. The paradigm of productivity became the dominant way of thinking in the mid-nineteenth century for the agricultural sector as a whole: farmers, financiers, and the emerging scientific community. However, accounting for the value of output relative to input was common practice earlier. Submissions to the British House of Lords Select Committee on Agriculture in 1821, for example, showed that wheat, yielding about £16 per acre, cost approximately £7.26 to grow, including the rental on the land. Other crops such as barley produced about £7.20, with much the same costs for input.

The emergence of the paradigm of productivity, beyond its application in the accounting of home finances and taxes, owes much to the contemporary state of the economies of Europe and the New Worlds. In about 1850, the real wages of European agricultural workers began to increase (expressed, for example, in terms of the number of kilograms of wheat earned per day), and the cost of renting land also increased. These trends continued until the end of the century. In

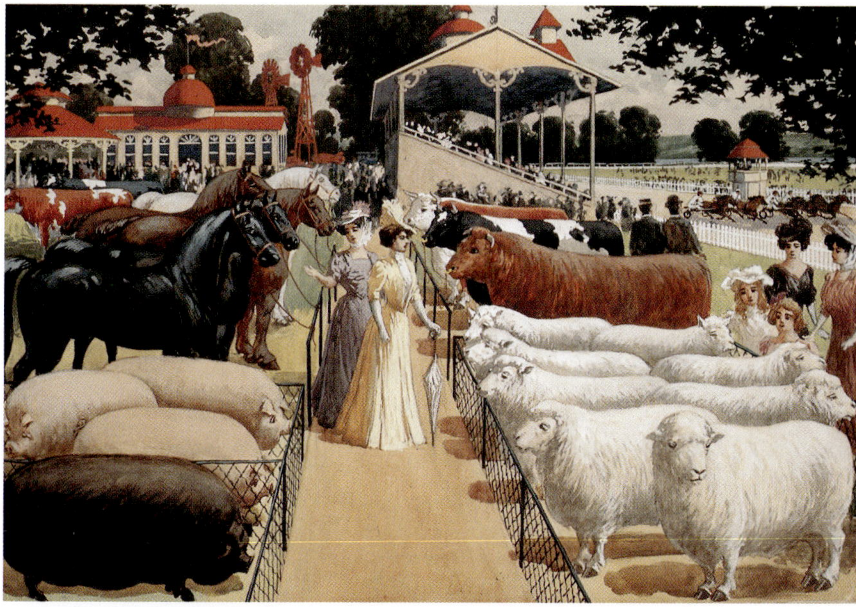

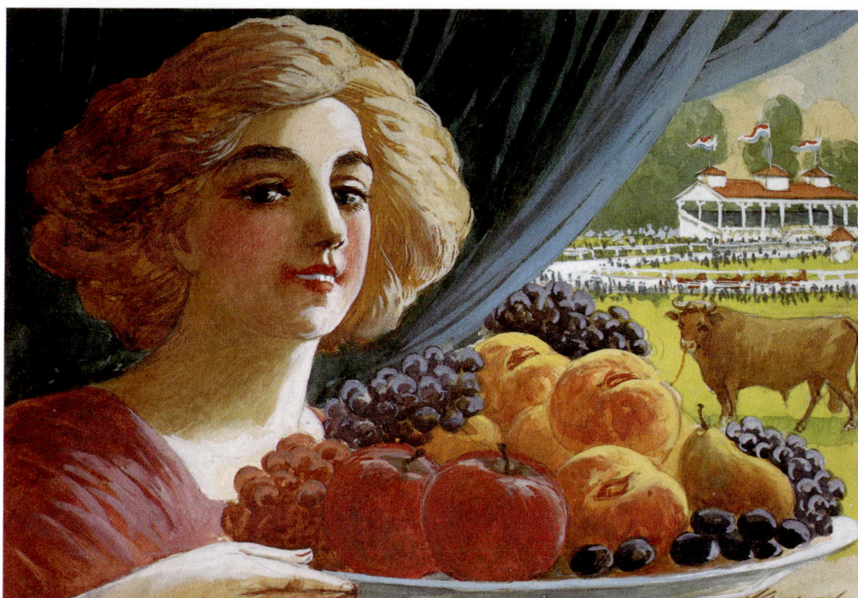

Figure 37
Advertisements for agricultural fairs. (Top) Artist unknown, *Avenue of Champions*, 1907 (opaque watercolour). With an avenue of champions in the foreground, this scene depicts the broad scope of fairs. Product displays, such as the windmill visible in the background, provided the opportunity to examine new technology. Domestic arts and lectures were presented in exhibition halls. Horse racing drew large crowds for exciting matches, and the best of the breeds were showcased through competition. (Bottom.) W. Layman, *Woman with Fruit (Fair Scene)*, circa 1910 (opaque watercolour). By 1813 – very early in America's fair history – Ladies' Day had been established to focus attention on women's contributions. The inclusion of women was important to ensuring widespread community participation.

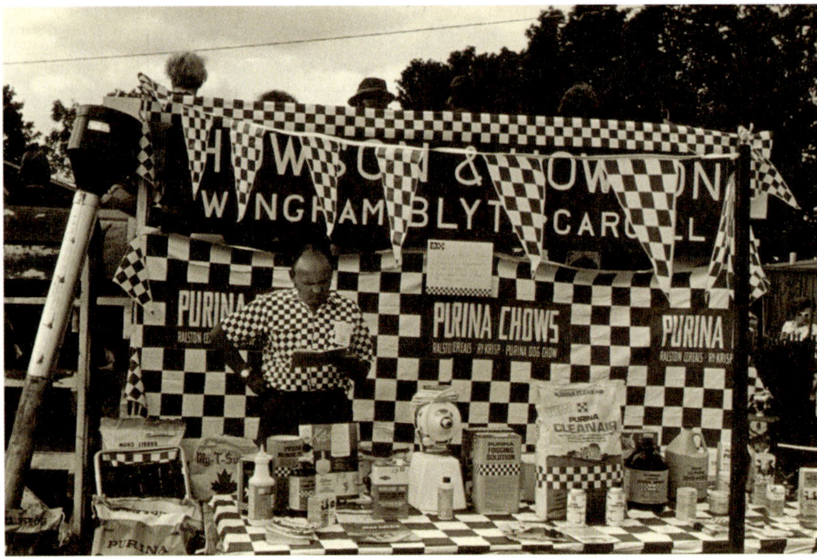

Figure 38
David Nasby, *Booth at Fall Fair, Wingham, Ontario, Canada*, 1971 (silver gelatin print).

1870 a day's labour could purchase 6.2 kilograms of wheat in Denmark, 7.5 in France, and 9.6 in Britain; by 1910, the same labour could purchase 20.3, 11.8, and 20.8 kilograms respectively.[5]

Similar trends were true for North America, Australia, and Argentina. During the period of expansion, land was cheap or free but labour was expensive (with the exception of convict labour in various Australian states and enslaved labour in the southern United States). Then, the cost of land and labour markedly increased until the late 1880s, when overcropping, drought, and economic depression hit agriculture. Economic and environmental storms during the 1890s led to the introduction of new technologies, advances in basic sciences, and an ongoing narrowing of the gap between the prices farmers received for their products (grain, meat) and their costs of production. The land was managed by people who could benefit personally if they could find better ways of agricultural production – the lords of the manors and tenants paying rent in England, and family farmers in the New World.

In nineteenth-century Britain, the "good farmer–bad farmer" movement arose as an initiative to improve agriculture through education beginning at the farm gate. Pamphlets were published illustrating the right and the wrong way of carrying out tasks such as building a manger, shearing sheep, or laying out a vegetable garden. A New World example of a self-help book is J. Russell Manning's *The Illustrated Stock Doctor and Livestock Encyclopedia*, first published in Saint Louis, Missouri in 1880, and subsequently issued by Hubbard Bros. in Philadelphia, Boston, Chicago, Cincinnati, and Atlanta, as well as in Guelph, Ontario, in 1882[6] (Fig. 39). This 1,082 page volume included topics from veterinary surgery to how to determine a cow's age by examining its teeth. The book also took a moralistic tone to encourage proper farm practices.

The quest for improved productivity is successful if innovation causes output to rise while input remains unchanged, or when output rises faster than the input costs, or when output prices fall more slowly than a corresponding fall in input costs. Success also requires that the extra value is retained, at least in part, by the farmer as well as being passed on to the consumer.

The paradigm of productivity in agriculture can be described in two different time frames. Rising output – higher yields per area – characterized agricultural innovation from 1850 to about 1950. Rising outputs with smaller increases in input costs have characterized the era from 1946 to the present day. The post-war availability of mechanization (especially tractors) and the introduction of pesticides changed the technology of farming through the period to 1960. The Green Revolution of the developing world occurred from 1960 to 2000. Our contemporary preoccupation with creating an industrial agriculture has the goal of reducing the value of outputs – cheaper, quality food for the consumer – while increasing profitability by reducing input costs.

Innovators abounded throughout the era of productivity. They comprised farmers, town-based entrepreneurs, enlightened amateurs, financiers, and scientists. As time passed, the core players changed from isolated individuals – gentlemen-farmers and scientists supported mainly by private benefactors – to "think tanks" of scientists and

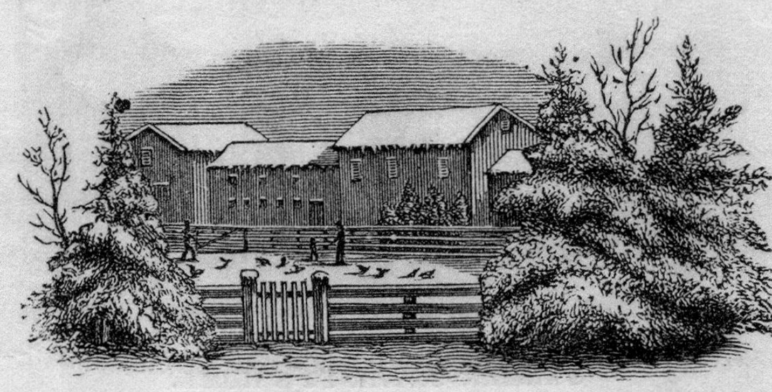

THE BARN OF THE PROVIDENT MASTER.

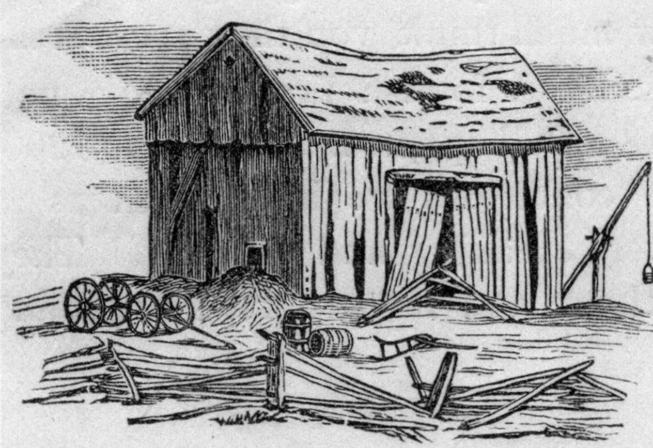

FARMER UNTHRIFT'S BARN.

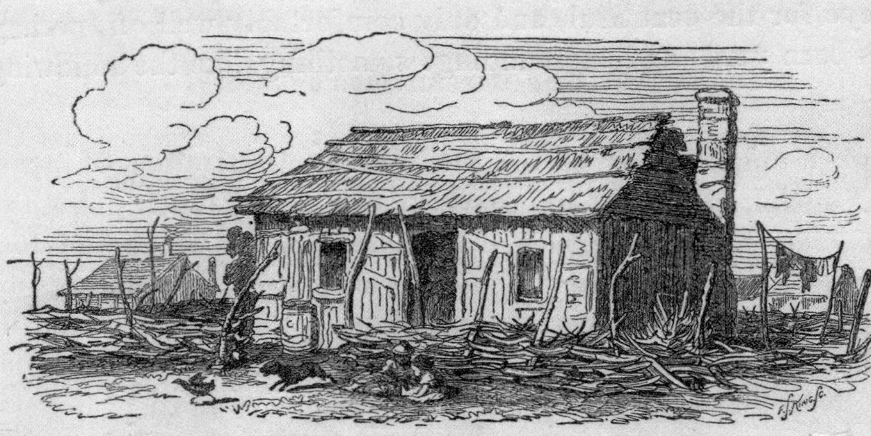

FARMER UNTHRIFT'S HOME.

product-development engineers supported by government and corporate money. They changed the technology and scope of agriculture through crop and animal domestication, by bringing a wider range of food crops and animals to market economies (illustrated by the different breeds of cattle and poultry that became available at this time), by developing technologies that enabled an expanded scale of operations, and by changing the way in which crops were grown or livestock reared.

This era of innovation contributed to the gradual compilation of theories and recorded experience that became the corpus of "agricultural science." A key person in the development of agriculture as a science, and the birth of what we now call agricultural chemistry, was Justus von Liebig (1803–1873). Von Liebig was labelled as "hopelessly useless" by his schoolmaster in Darmstadt, Germany. Nonetheless, perhaps driven by his father, who was a dealer in chemical supplies, von Liebig pioneered an analytical approach to chemistry, mostly as a professor at the University of Munich, where he worked from 1852 until his death. In 1840 he defined the "law of the minimum," by which the nutrient that is in shortest supply imposes the

Opposite: Figure 39
Manning illustrates how the labour of "good farmer" and "bad farmer" bears different fruit. "The surroundings of a man in any condition in life, whether he be rich or poor, are an index to his character. The animus of all men is to make money, but some possess in connection, a love of the beautiful. Without method in labor no man can be successful. The farmer who has method, and an eye for the beautiful, and only comfortably well off, perhaps, will show his barn yards and surroundings something like [this] illustration [top]. His barns are tight and ample, and filled to the ridge-pole with fodder. His yards are protected with shelter-belts and wind-breaks, his pastures and meadows ample and luxuriant, and his crops well tilled and heavy. Inside his barns will be found a place for everything and everything in its place. On the other hand we give a view of the barn of the improvident master [centre]. His well, simply a hole in the ground where the drainage of the yard may enter, the roof of the hovel rent and torn, the dilapidated doors propped up with rails, the weather-boards fallen or falling off, and the whole thing shaky, like the master's mind. Fine stock, fat, and well groomed, have come out of hovels of barns; they were made warm and comfortable. It is not the most expensive structures that always contain the best stock, but in the end the better barn will be built. We have never seen good stock issue from such a barn as we have shown, and it only needs to show the house the farmer lives in [bottom], with its brush heap, its line of ragged clothes, the ragged, dirty children, and generally dilapidated appearance, to complete the pictorial story of general unthrift." From *The Illustrated Stock Doctor and Livestock Encyclopedia, pp. 149–50.*

greatest limitation on crop yield. Von Liebig's key publications in 1840 and 1855 built on work by the German botanist Carl Sprengel (1787–1859) to such an extent that some view Sprengel as the co-founder of agricultural chemistry and, specifically, as the first to formulate the "law of the minimum." Wherever the credit lies, the importance of this discovery cannot be overestimated: insight into a single, limiting factor on productivity opened the door to 150 years of research, which continues today, into factors that impede productivity in particular situations and how they can be overcome.

The first and arguably most famous agriculturists to apply the law of the minimum to the science of plant nutrition was an unlikely pair, John Lawes and Henry Gilbert. Lawes and Gilbert formed an extraordinary partnership in the history of agriculture and exemplified the nineteenth-century scientist-innovator. John Bennet Lawes (1814–1900) inherited his family's estate, Rothamsted, near St. Albans in Hertfordshire. In his day, Rothamsted would have looked much like any other estate with its dense hedgerows and coppices, ornate gates and gate-houses, and large fields shaved of crops in late autumn. Today, Luton airport is twenty kilometres away, and nearby the dreaded M25 motorway orbits north London. On the estate is a two-storey, rather institutional-looking, building aligned with a row of golden-leafed oaks. Behind are bog-red brick buildings, two and three storeys high, resembling barracks rather than structures befitting a country estate. The main building is situated between modest cottages and two-storey townhouses, for the local village has overtaken this section of countryside. In front of the main building is a sizeable oak, planted in 1993 by the Royal Agricultural Society to mark the 150th anniversary of the start of Lawes' agricultural experiments.

Lawes was a gentleman farmer with an inquiring mind. He noticed that spreading bones as fertilizer on fields, a practice common in the early 1800s, had different effects depending on the field. He converted a bedroom into a laboratory, ground up the bones, and treated them with sulphuric acid. In 1842 he patented the result – "superphosphate" – and then spent the next thirty years building a fertilizer business. In 1872, he sold his successful factories, which allowed him to direct all of his considerable energy to experimentation at Rothamsted.

Joseph Henry Gilbert (1817–1901), by contrast, was born into genteel poverty. He was the son of a Congregational minister, and his aunt was known as the writer of "Twinkle, Twinkle Little Star." His image as passed down through the generations was of a rather severe individual who was partly blind and quite unlike the benevolent squire, Lawes. Nonetheless, it was Gilbert who was trained formally as a chemist, taking a doctorate with the eminent Justus von Liebig. He became Lawes' assistant, and with his solid scientific background was quickly promoted to junior partner in 1843. Gilbert, the thorough experimentalist, travelled widely and was probably not all that fearsome. As well as working at Rothamsted, he was professor of rural economy at Oxford University from 1884 to 1890.

Lawes and Gilbert began in 1843 by planting wheat fields and experimenting with all sorts of fertilizers and crop-management practices, including the use of sewage sludge. (Interestingly, these same fields have been continuously cropped to the present day.) In 1855, the report of their experiments on continuous cropping in the Broadbalk fields established the importance of nitrogen as a nutrient for wheat. These experiments were followed by a long series of trials on the value of legumes such as clover. In a series of trials, the best of which Gilbert reported in a paper read at Montreal in 1882 (Gilbert travelled and lectured widely), they showed that clovers collected nitrogen from the soil.[7] This work ended a thirty-year public controversy with Gilbert's mentor, Liebig, who disparaged the sort of field experimentation to which the Englishmen had dedicated their lives. The basis of the slanging match was whether plants obtained their nitrogen from the air (as proposed by Liebig), or from the soil, and were thus amenable to stimulation by adding fertilizer (as defended by Lawes and Gilbert). In fact, both were correct: in the general case, it is taken up from the soil, although in legumes nitrogen is largely, but not exclusively, taken up through the fixation of nitrogen from the atmosphere. The fixation of atmospheric nitrogen by bacteria on the roots of legumes was proposed in 1886 by the German botanist Hermann Hellriegel (1831–1895). To their credit, this caused the duo immediately to begin further experimentation that verified Hellriegel's theory. Throughout their work, Lawes and Gilbert attempted to calculate nutrient budgets: how much fertilizer was taken up by crops in comparison with what was available in the soil or applied as fertilizer. Their partnership, productive until they were both old men, lasted fifty-seven years. Although the march of agricultural productivity from 1850 to 2000 was due to a great number of farmers and scientists, the original and prodigious work on plant and animal nutrition, fertilizers, and crop rotation by Lawes and Gilbert has no equal.

Gregor Mendel (1822–1884) was a contemporary of the English fertilizer-and-husbandry duo, who undoubtedly would have found him an interesting dinner companion. He was ordained a priest in the Augustinian abbey of St. Thomas at Brünn, Moravia (now Brno, Czech Republic) in 1847, but in 1850 failed the biology examination required for certification as a teacher. Nonetheless, he spent seven years painstakingly hybridizing thirty-four types of pea plants, noting the shape of the seed, the length of the stems, and different flower colours of the progeny.

Mendel sketched his ideas on the back cover of an 1849 book on hybridization and did not have the opportunity to read the German translation of Darwin's *On the Origin of Species* (1859) until the year he finished his analysis. However, he extended Darwin's work significantly by showing that some traits are passed from parent to offspring in simple, predictable numbers: the quantitative basis of inheritance. Mendel presented his results to the Brünn Natural History Society in 1865 and published a paper on "Experiments with Plant Hybrids" through that society in 1866.[8] It would be satisfying to relate that Mendel had a major influence through his work on quantitative genetics; however, this was not so. He failed to replicate his published findings when he tried to unravel the inheritance pattern of flower colour in hawkweed. The rest of his career was consumed with administrative work.

Other scientists, unaware that Mendel had shown that genetic traits could be quantitatively inherited, also struggled to define the quantitative basis of inheritance. Among them were the Australian William Farrer (1845–1906), who was breeding wheat, and American Charles Quinby, who was creating dwarf sorghums. Mendel's work was rediscovered in 1900. Then, belatedly, his pioneering work led other generations of scientists to search for the building blocks that create the simple ratios of inherited traits that he discovered. This did not happen until 1956, when James Watson and Francis Crick published their model of the structure of DNA. The rest is the recent and rapid history of molecular genetics. Today, Mendel is honoured where he spent all his working life, in Brno, by a larger-than-life statue. It stands in front of the church and white and terracotta-tiled monastery (now leased as business offices) where he did his research. Appropriately, the statue overlooks the small park that was the site of the glasshouse where he grew his peas.

Thousands of plant and animal selectors and breeders contributed to early agricultural research. Plant breeders, selecting high-yielding genotypes or manually cross-pollinating lines, created successive generations of crops that had higher yielding potential or that, just as importantly, would not be susceptible to diseases such as rust. The outcome of their work, as with better management practices and fertilizers and the ruthless control of weeds, has been to raise the yield and reliability of crops. In the 1850s the common crops yielded about ten to fifteen grains for each seed sown, and distributed about 10 per cent of their growth into the inflorescence or seed-head. By the 1980s the ratio of seed yield had increased. The harvest index, the ratio of weight of seed-head to the rest of the plant, had increased from about 0.1 to about 0.55 by 1980.

Sophisticated management, the tailoring of crops to particular regions, large-scale specialized machinery, and the economies of scale and simplicity, are all reasons why agriculture has become progressively more specialized and less diverse within any particular region. This has changed the way we think about agricultural landscapes. At the same time, the combination of progressive increases in crop yields, specialization, and more one-dimensional landscapes has given rise to changes in the character of rural towns. Perhaps one of the most enduring symbols of the paradigm of productivity is the North American grain elevator, whose counterparts dominate rural towns in all of the crop-growing regions of the world. These columnar giants were a favourite subject of painters and photographers in the 1920s and 1930s: Charles Demuth's classic grain elevator (Fig. 40) expressed the belief in the power of machines to bring progress

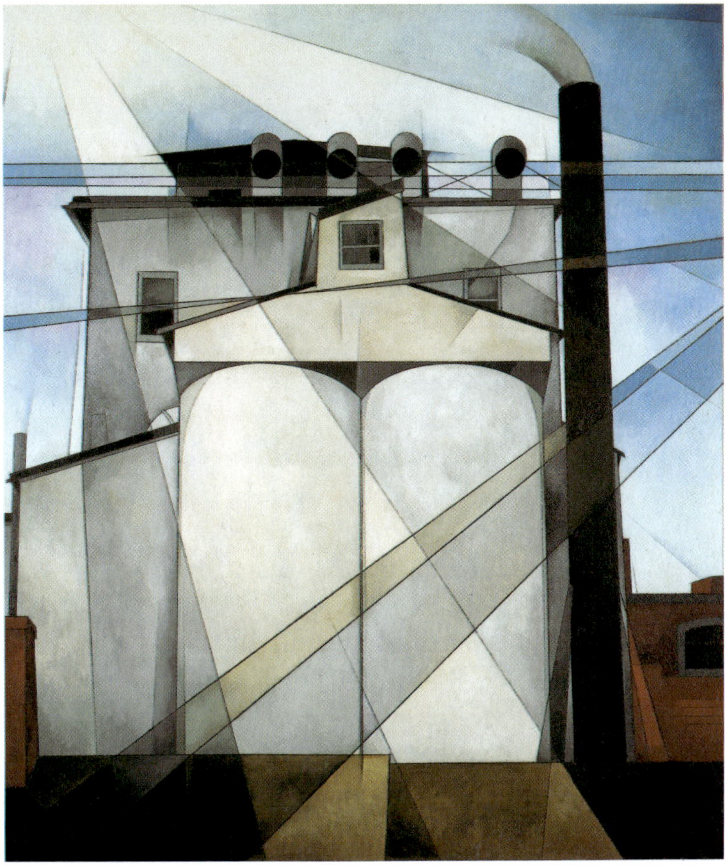

Figure 40
Charles Demuth, *My Egypt*, 1927 (oil and graphite on composition board). The title of Demuth's masterpiece likens the colossal structure of the grain elevator to the nobility and powerful geometry of the Egyptian pyramids. In his depictions of industrial buildings and skyscapers, Demuth expressed a prevailing belief in the power of the machine to effect progress and prosperity. The model for this elevator is in Lancaster, Pennsylvania, the artist's hometown, to which he was confined as a bedridden invalid in mid-life.

and prosperity. As in art, so it was in agriculture, with faith in the paradigm of productivity, at least for the mainstream of the industry until the 1960s or 1970s.

The advent of the tractor had a dramatic impact by lessening the need for manual labour in the fields, increasing the area that could be cultivated, and providing a vehicle to take produce to markets. Mechanization also increased the reliability of the food supply by virtue of its speed: crops could be planted and harvested closer to the optimal times. At the same time, mechanized agriculture broadened the division between rich and poor, since only the wealthy could afford to buy machines. The tractor, while introduced in most Western economies in the 1930s as steam-powered machines and advocated through government-sponsored "tractor schools," did not have widespread impact until reliable and relatively cheap petrol-powered machines became popular after World War II (Figs. 41, 42).

World War II itself had a big impact on agriculture, particularly in Europe. Armed occupation, isolation and the cutting of supply lines heightened awareness of the need for "food security:" self-sufficiency in staple foods and fibres. Britain was a most striking example. It was fighting a war against a former trading partner (German–British trade amounted to about four per cent of British imports and exports) and became quickly isolated from occupied Europe. This naturally shifted emphasis to North America and the southern hemisphere as sources of food imports, which in turn were squeezed by the German U-boat campaign to cut supply lines. Relative to 1938, the volume of food imports to the United Kingdom had fallen to 60 per cent by 1942. Beef, veal, and butter fell to about a third. This had two effects: (1) a rising consciousness of food security; and (2) the implementation of government campaigns, willingly taken up, to grow more food, mostly in vacant lots, parks and city plots. Thus began a national movement into "urban agriculture." The British "Dig for Victory" campaign was begun at the outbreak of war, and it was estimated that 1.4 million people had allotments – small plots for growing household food – at its peak. Kensington Gardens, golf and tennis clubs, window boxes, and even the moat of the Tower of London were turned over to hand-hoes and vegetable patches.

The disruption of trade and strangled supply lines spurred rationing. In Britain, rationing was introduced on 8 January 1940, limiting the consumption of bacon, ham, sugar, and butter. By June of 1941, each person was limited to one fresh egg per week – if it was available – and meat had been cut progressively to only five pence' worth per week. Some relief, albeit with an unusual taste, was introduced with dried egg powder in June 1942, but even this was rationed. The rationing of eggs, along with many other "basic" foods, continued until March 1953.

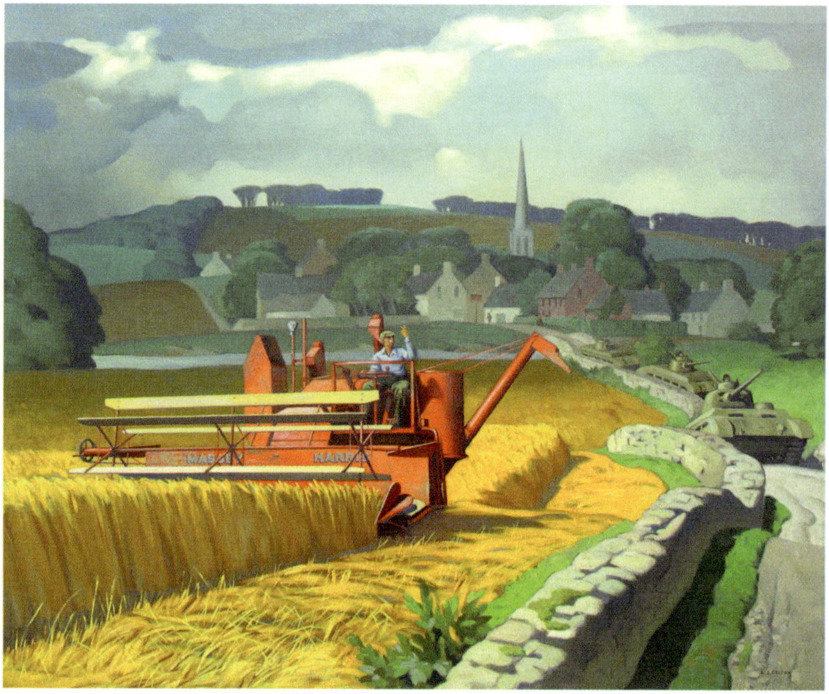

Figure 41
A. J. Casson, *"V" is for Victory*, circa 1939–45 (oil on masonite). This World War II era painting was commissioned by the Massey Ferguson Company to use as a magazine advertisement. In an idyllic landscape a delighted farmer drives his new combine front and centre in the composition. The setting is farmland on the outskirts of a European town; the farmer waves to soldiers in approaching tanks. The advertisement acknowledges the significant role of farmers in the war effort. For potential customers the portrayal of rich farmland on a bright sunny day adds to the allure of owning new farm machinery.

The "Dig for Victory" campaign in the United Kingdom, community-initiated urban agriculture in occupied continental Europe, and rationing throughout much of the world, turned attention back to the need for production: each country wanted to independently produce the food it needed, with little regard for cost or efficiency. This was reinforced by the wearying post-war years when rationing was retained irrespective of victory or defeat.

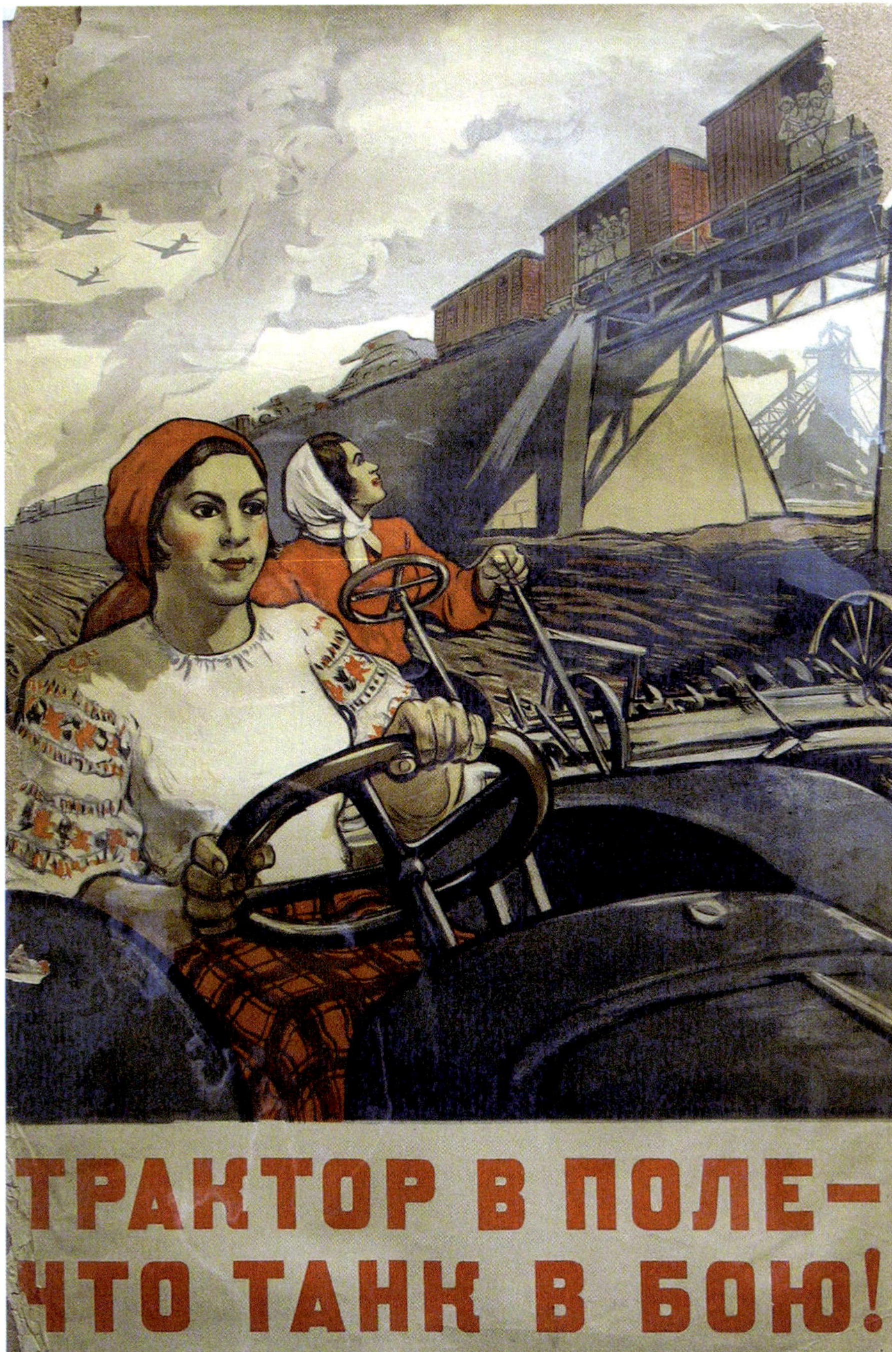

Figure 42
Victor Ivanov and Olga Burova, *A Tractor in the Field Is Worth a Tank in Battle*, 1942 (propaganda poster). This illustration acknowledges the vital role women played operating farms during World War II.

After World War II, heightened concern about food security led European nations to adopt economic policies that progressively introduced favourable long-term government price support for agriculture. There was also a growing humanitarian concern for impoverished countries that faced food shortages. Without increased agricultural production, many millions in these countries would have starved. By solving problems of food shortages, wealthier countries were provided with opportunities for treating a food exporting market.

Thus, domestic concerns about security, humanitarian needs, and global interest in trade created an ideological environment, particularly in Europe, that renewed consideration of production methods: the quantity of food produced per country regardless of the efficiency of production. Time trends for Britain and the Netherlands bear this out: wheat yields and national production increased, as did inputs such as fertilizers, from the 1940s to about 1985 (Appendix 2).

Driven by an enthusiasm for food security, war technology was quickly employed to transform post-war agriculture. Mass production technologies, directed to tanks and personnel carriers, shifted to tractors. Chemical production technologies turned to creating world-wide availability of inorganic fertilizers (following Lawes and Gilbert) and pesticides. Pesticides were introduced progressively as new formulations became available and were shown to be selectively effective against target weeds and insects. Over decades, some were withdrawn because they became ineffective. Their repeated application created an artificial pressure on the target populations so that only those weeds or insects that were genetically different from the bulk of the population and unaffected by the pesticide would survive and multiply. Chemicals were withdrawn because they caused a shift in the population in favour of resistant strains, thus making the pesticides ineffective, or because of concerns about human or environmental safety.

The biggest "bang" from pesticides use was obtained from DDT (dichlorodiphenyltrichloroethane). This compound was originally synthesized in 1873, but it was not until 1939 that Paul Hermann Müller (1899–1965), while working at the chemical manufacturer J.R. Geigy in Basel, Switzerland, discovered its utility as an insecticide. The chemical was quickly picked up and used to delouse troops during the war. Müller won the Nobel Prize for Physiology or Medicine in 1948 for his discovery. DDT was by then used widely in agriculture, as well as more generally to kill mosquitos, a vector for malaria, and snails, which could carry leishmaniasis. However, by the early 1950s there was ample evidence that DDT, in addition to killing arthropods indiscriminately, was highly toxic to fish. Further, it was not broken down quickly but instead was deposited in fatty tissue so that it accumulated in animals and humans throughout their lifetime.

In 1962 the American ecologist Rachel Carson (1907–1964) published her influential *Silent Spring*,[9] bringing public attention to the dangers of widespread use of DDT. She was initially dismissed as an amateur and alarmist, rather peculiar charges given her Master's degree from Johns Hopkins University and her position as editor-in-chief for the US Fish and Wildlife Service. Nonetheless, public interest was engaged. Public concern and the availability of alternative chemicals, at least for some targets, resulted in a ban on the use of DDT in the United States beginning in 1973. In total, 675 thousand tonnes had been sprayed in the United States. The chemical is now almost washed out of the environment in North America, through breakdown and shifting with sediments to the sea over the past twenty-five years.

In 2000, the United Nations Environment Program achieved an important international agreement for the worldwide ban of twelve persistent organic pollutants. DDT was listed as one of the most dangerous; however, it alone was singled out for some continued use in low-income countries, where there is currently no low-cost alternative to kill mosquitos. The reduction or elimination of DDT in parts of Africa through the 1980s and 1990s resulted in a resurgence of malaria and leishmaniasis.

The simple approach of repeatedly spraying pesticides on a crop or applying it as a drench or spray on livestock until all of the target pests are eradicated evolved in the 1960s. Later this evolved into a philosophy of "integrated pest management," known as "IPM" in the trade. IPM was developed by entomologists and dates to a landmark publication in 1959[10] that proposed that farmers should not only tolerate, but aim to manage, ongoing populations of pests below certain threshold values. The authors, Stern, Smith, Bosch, and Hagen first described the population density of the Colorado potato beetle prior to about 1850 and how it had jumped to much higher levels following the widespread culture of potatoes in the United States. They also pointed to other causes for increases in pest populations, such as when pests are transported to new agricultural areas where their natural predators and diseases do not exist and consequently multiply without check. At the same time, there was ever-growing demand for greater control of pests because consumers became increasingly intolerant of imperfect produce. Consumers of fresh and frozen vegetables today do not tolerate pest damage, and either request discounted pricing or will not buy the damaged food at all, although it may still be quite nutritious and safe to eat. Having set the scene, the authors proposed a new answer: instead of continuing to try to eliminate pests, a better solution is to mix biological and chemical control to maintain the pests at acceptable levels. The maintenance of these sub-threshold populations can be achieved with chemical applications that kill some pests while letting other, beneficial, organisms, survive and eat the "bad guys;" this approach is less harmful to weeds and insects that might be beneficial and is less harmful to the environment generally. In 1968, IPM was defined by the United Nations Food and Agriculture Organization (FAO) as:

> a pest management system that, in the context of the associated environment and the population dynamics of the pest species, utilizes all suitable techniques and methods in as compatible a manner as possible and maintains the pest populations at levels below those causing economic injury ... It is not simply the juxtaposition or superimposition of two control techniques (such as chemical and biological controls) but the integration of all suitable management techniques with the natural regulating and limiting elements of the environment.[11]

In the 1980s, the overuse of fertilizers and indiscriminate use of pesticides caused the non-farming public to say "enough!" Legislation was introduced in Europe to curb the use of fertilizers and pesticides and to make farming more sustainable. A paradigm of responsible productivity surfaced, followed by a paradigm of sustainability. These are topics for later chapters.

There are numerous areas in the world where people have resisted, ignored, or been unable to adopt the productivity paradigm. In many of these places, both traditional farming practices and biodiversity have been retained. Landscapes have not changed; nor have ways of life. The downside of not improving husbandry and plant and animal breeding is stasis. In cases where the human population has increased and the productivity of local agriculture has not kept pace, the outcomes are out-migration or famine. One example is the cultivation of teff in the highlands of Ethiopia. Since before recorded time, farmers in this region have favoured this grain, which today retains poor seedling vigour and therefore does not compete well with weeds. Its low harvest index is comparable to that of nineteenth-century European wheat. Teff looks like a weedy grass about thirty to forty centimetres tall and has a floppy head that contains seeds as fine as grains of sand. It is harvested by hand, and the seed is ground up and mixed with water and milk to form a watery mash that is allowed to ferment and then poured onto a round hot plate over an open fire. The resultant flatbread is airy and sour, almost effervescent to taste, the product of a traditional agriculture and way of life that has been passed by, outside the march toward productivity and uniformity.

While Western agriculture began to rely more heavily on fertilizers and pesticides through the 1950s and 1960s, and yields were increasing more than fast enough to keep pace with population growth, concerns were mounting about global population growth, particularly in low-income countries. The concern proved to be well founded: the world's population grew from 2.5 billion in 1950 to 6.1 billion in 2000. In the 1950s, governments and development agencies (specifically, the Ford Foundation and the Rockefeller Foundation) identified two problems that could impoverish the world: stagnant food production and population growth. However, to identify these issues is one thing; to address them requires leadership and creativity in transplanting the scientific basis for the productivity of Western agriculture to the particular conditions of low-income countries.

This transplantation of agricultural practice became known as the "Green Revolution," a term coined in 1968 by William Gaud, former director of the United States Agency for International Development (USAID). The revolution unfolded from 1960 to about 2000; some argue that it is ongoing, others that it has lost momentum. The Green Revolution was planned and orchestrated by government-supported scientists working to improve the productivity of farmers in low-income countries not only to produce higher yields of food for their families but also to create, with surplus production, more cash or barter power to obtain household and farm items. The introduction of high-input technologies and new crop varieties developed by a network of international research institutes collaborating with national (country-based) centres and extension programs resulted in increasing crop yields in developing regions, particularly Asia (Fig. 43).

The leadership of two administrators, both born in small-town America, was crucial to the Green Revolution. Jacob George Harrar (1906–1982) was an avid naturalist who, after the completion of his first university degree, worked in Puerto Rico before returning to the United States to obtain a doctorate in plant pathology. He began working for the Rockefeller Foundation in 1943, first in Mexico and then in New York, succeeding Dean Rusk as the foundation's president. Robert Chandler Jr. (1907–1999) started as a horticulturist in Maine and worked with Harrar in Mexico before becoming president of the University of New Hampshire in 1950. Five years later he moved back to the Rockefeller Foundation, first working with Harrar in New York and then moving to Asia to build the first institute that they created.

Harrar had moved to Mexico in 1943 to lead a new – and at that time unique – collaboration between the Mexican government and the Rockefeller Foundation, which aimed to improve Mexican crop yields. On 11 September 1950, Placido Mapa, Secretary of Agriculture and Natural Resources in the Philippines, wrote to John D. Rockefeller III, inviting the foundation to "start a project here com-

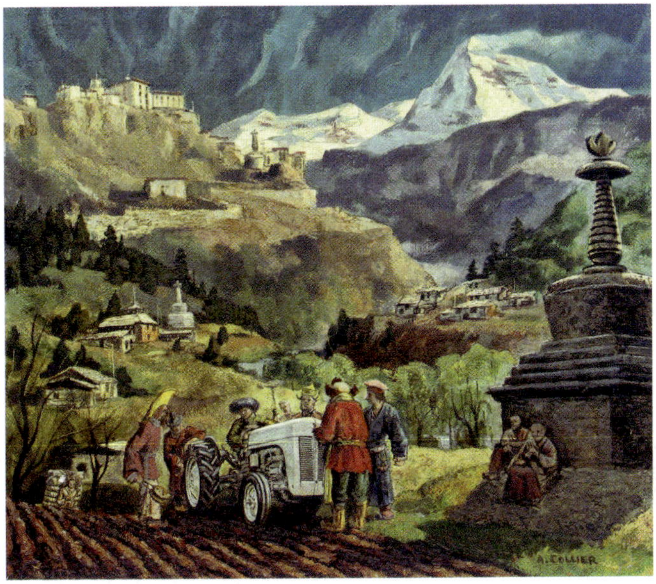

Figure 43
Alan Caswell Collier, *Ploughing in Tibet*, not dated (oil on canvas board). Delighted farmers examine a tractor in Tibet. This painting is not unlike A.J. Casson's illustration for the Massey Ferguson Company (Fig. 41).

parable to the corn project in Mexico." By 1958, Harrar and Chandler had scoped the idea of an institute for rice improvement in the Philippines that "would help solve many of the problems of rice production in the tropics."[12] Harrar proposed:

> One approach to this objective might be the establishment of an International Rice Research Institute:
>
> A) To bring together available information on rice and its management,
>
> B) To recruit and organize a group of competent resident investigators who will work on the basic and applied problems of rice production, and
>
> C) To establish international cooperative relations directed towards increasing the effectiveness of research and its general application.[13]

The concept was approved on consecutive days in June 1959 by the Rockefeller and Ford foundations, and Chandler drafted the articles of incorporation by November. By March of the following year he was in the Philippines, working as director of the International Rice Research Institute (IRRI). A photograph of the first meeting of the IRRI Board of Trustees in April 1960 in Manila shows eleven non-Philippino trustees, including Harrar with a long, tired-looking face, and Chandler, chubby, dark-haired, and bow-tied. Interestingly, the Philippine Congress approved the setting up of the institute in a rushed evening sitting on 19 May 1960, the last of the legislative year; it received only one paragraph in the day's newspaper, which was mostly concerned with the passage of the national budget.

By October 1961 Chandler had recruited 13 senior scientists, including Ben Vergara and Bienvenito Juliano from the Philippines, and others from the United States, China, Sri Lanka, and Japan. They began work on the project in the following year. The eighth rice hybrid made at IRRI in 1962 was eventually commercialized as IR8 in 1965. IR8 made a major impact on increasing rice yields throughout Asia and would alone have been enough to constitute a "Green Revolution."

Meanwhile, the collaboration between the government of Mexico and the Rockefeller Foundation, aimed at improving yields (particularly of maize), was formalized in 1963 as the Centro Internacional de Mejoramiento de Maíz y Trigo (CIMMYT) or the International Maize and Wheat Improvement Center. Norman Borlaug, the head of the agriculture program in Mexico for the Rockefeller Foundation, became head of plant breeding at CIMMYT when it was formed and led the program until his retirement in 1979. However, the administration of the cooperative partnership was, at least in contrast to IRRI, tangled and slow-moving, and beset by confusion as to who was responsible for what and by student protests about Yankee imperialism in Mexico. There was no doubt about the gravity of the issue: a memorandum from this period between the Mexican government and the Rockefeller Foundation begins: "World population is increasing at an explosive rate. Unless world food production can be accelerated ... a world famine is inevitable in the 1970s."[14]

Years of excruciating negotiations and the drafting of more formal arrangements led to a partnership agreement between the Mexican government and the Rockefeller Foundation in April 1966 and the first meeting of the board of the newly restructured and formalised CIMMYT in September. The paradigm of productivity was front and centre in the purpose statement from that meeting: "To assist nations throughout the world to increase the production of wheat and maize."[15]

Following the lead of the IRRI and CIMMYT, other governments and charitable organizations became supportive of the concept of international centres, and the Rockefeller and Ford foundations recognized that they could not, alone, fund a network capable of addressing the problem of global food shortages. In April 1969, Harrar hosted officials from fifteen national agencies at a workshop at the Rockefeller Centre at Bellagio, Italy. The Centre, a villa on a promontory that has grown grapes since Roman occupation, overlooks both arms of Lake Como. Over three days the group agreed on the need to quadruple world agricultural output within thirty-five to forty years. After three more meetings, the group agreed to support the established institutes and explore the founding of others to serve other needs and regions. They also agreed to set up a more permanent body of donor agencies: the Consultative Group on International Agricultural Research (CGIAR) was formed in 1971 as a group of "28 organizations with a declared interest in helping to stimulate the agricultural sectors of the developing world."[16] CGIAR progressively set up an international network of sixteen research centres whose goal was to improve productivity. Other centres have focused on other crops and livestock, and on particular agro-ecological regions. The centres have contributed to the release of over eight thousand crop varieties in one hundred countries, increasingly working in conjunction with national agricultural research centres.

An analysis of the productivity gains achieved by the Green Revolution indicates that it has made a positive impact in all areas except sub-Saharan Africa, where it has achieved only very marginal gains.[17] Without the international research that created the Green Revolution, modelling suggests, crop production would have ended up in 2000 between 14 and 19 per cent lower in the developing world, although to maintain food supply even at this level the agricultural area would have expanded more than it did in reality, likely with attendant environmental problems:

> The model indicate(s), that in the absence of international research, the world would have experienced a "human welfare" crisis. Caloric intake per capita in the developing world would have been 13.3 to 14.4% lower, and the proportion of children malnourished would have been from 6.1 to 7.9% higher. Put in perspective, this suggests that the Green Revolution (between 1961 and 2000) succeeded in raising the health status of 32 to 42 million preschool children. Infant and child mortality would have been considerably higher in developing countries as well.[18]

In his acceptance address for the Nobel Peace Prize in 1970, Borlaug put it succinctly: the Green Revolution, which even then was the target of criticism, had at least contributed through improved crop productivity to averting famine and feeding a substantial proportion of the growing human population; once people were no longer starving they would be in a better position to contemplate issues such as environmental sustainability.

Although fifty years of crop breeding has clearly had an impact – an estimated thirty to forty million preschool children escaping malnutrition – it is also noteworthy that it has changed the technologies and landscapes of the world. Think of traditional images of agrarian life in Asia, showing men with hoes, and oxen or water buffalo ploughing the rice fields. The outcome of the pursuit of productivity and food security is the introduction of hand-operated rotary hoes and tractors, making scenes such as the one depicted in Figure 44 increasingly rare. Likewise, consider the art and photographs that depict women ankle-deep in water, transplanting rice seedlings: a practice that, having lasted for millennia, will die out in our lifetime. The practice of giving the slow-growing seedlings a "head start" in seedling beds before being exposed to the harsher field environment is no longer needed because of the improved vigour of modern seedlings. The beneficiaries are women – transplanting is

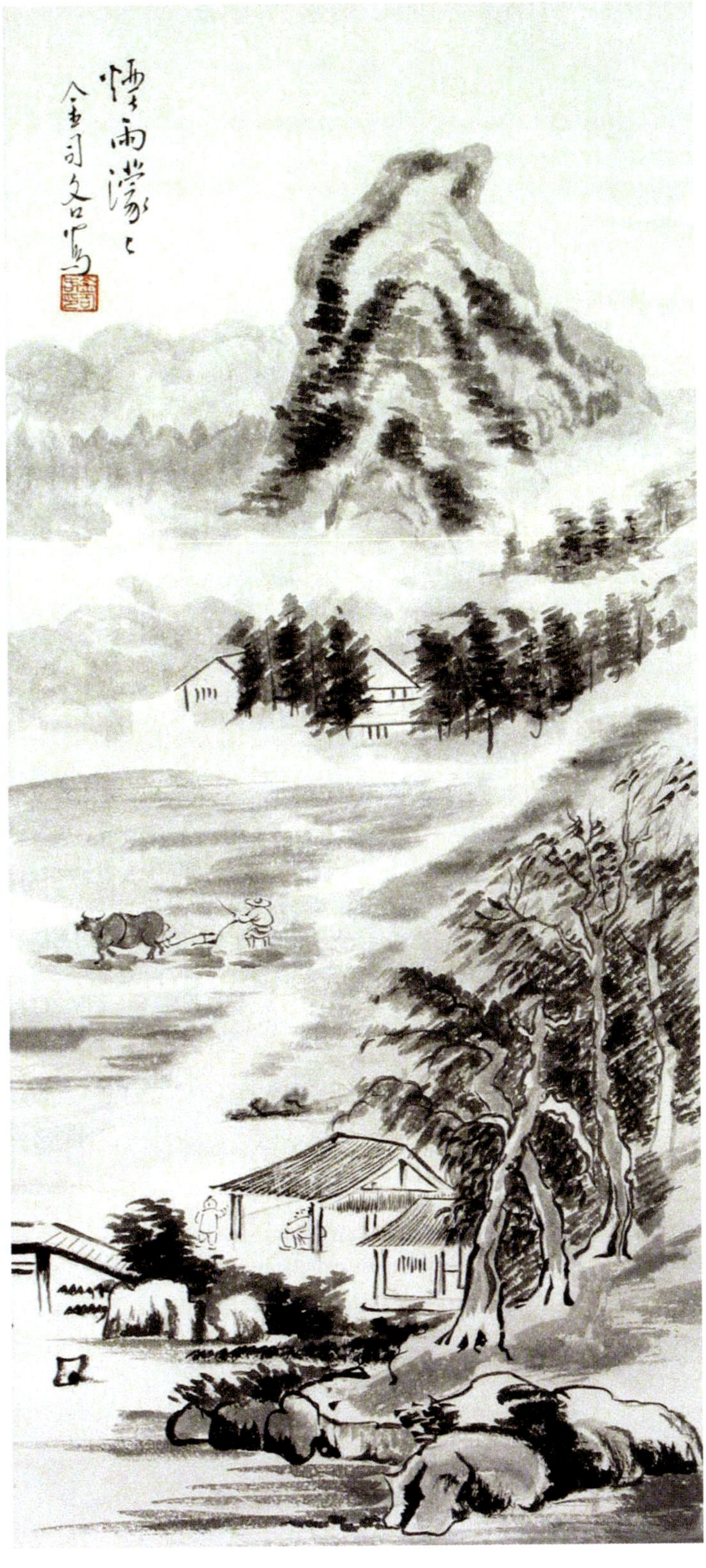

Figure 44
Anthony Kingscote, *Ploughing the Rice Paddy*, circa 1968 (ink and wash on paper). Executed in the traditional Chinese brush-painting style by a Canadian artist who learned the technique while living in Manila, this painting romanticizes traditional rice farming. To prepare for planting the farmer guides a simple plough pulled by a water buffalo through the flooded field.

tedious, backbreaking work – and their household economies, for women can now work for wages off the farm, when such work is available.

Another agricultural practice that has been transformed by the Green Revolution is the procedure for threshing grain. The traditional method was to bring the heads of the rice or wheat attached to their stalks to the threshing floor, a central point in the field, or next to the settlement. Here on a stamped-down patch of bare ground women would beat the grain from the heads, or oxen would trample it in the sun, knocking the grain out onto the dirt to be gathered into bags. As soon as high-yielding crop varieties were introduced, the volume of grain and work made this traditional, social task very burdensome, especially as more workers sought off-farm employment and the increasing labour of the harvest fell on fewer people. Small threshing machines were introduced, reducing labour and waste.

The "next generation" of the green revolution is under way. While continuing to chase marginal increases in grain yield, the current generation of plant breeding seeks to make quantum improvements in the nutritional value of the grain. An example is the improvement in the nutritional value of rice with the development of "golden rice" types, which contain a higher content of the metabolic precursors of vitamin A. Varieties released in 2005 have twenty-three times more beta-carotene than traditional varieties.[19] This should help to overcome vitamin A deficiency and will be particularly valuable for children in countries such as India, where the incidence of vitamin A deficiency is as high as sixty per cent.

Less attractive effects of the pursuit of productivity have been the adoption of crop cultures that are more reliant on herbicides and pesticides than traditional varieties. These have two effects: reduced insect and animal activity make the soil less textured and porous; as a result, water is more likely to run off the surface and erode the soil. Indirectly, the Green Revolution has increased the cost of cropping, which has made many small farmers and tenants more, not less, dependent on lenders to finance their crops. Although this might not be problematic if productivity – and profit – has increased, dependency on financial support makes farmers more liable to lose their livelihood if the crop fails: bigger inputs mean bigger failure, when it occurs.

The intensification of farming during the Green Revolution was one important aspect of the transformation of agriculture that unfolded during the period dominated by the paradigm of productivity, from about 1850 to 2000. A second important aspect was the extensification of agriculture: the clearing of forests, usually in marginal lands, to provide new land for cultivation. This era of expansion is described in the last chapter. In some countries, productivity and expansion have evolved concurrently. To bring these two threads together for a moment, it is worth noting the increase in total food production in various low-income countries from 1960 to 1980, and then from 1980 to 2000, described in Everson and Gollin's analysis.[20] In Asia relatively small increases in area under cultivation were achieved before 1980 and virtually none since. The additional food production in Asia during the Green Revolution has arisen almost wholly from gains in productivity. In Latin America, expansion played a role equal to productivity in increasing food supply before 1980. In the period 1980–2000 all the gains in countries other than Brazil resulted from increased productivity. In sub-Saharan Africa, both expansion and productivity were important until 1980, but while expansion continued in more marginal areas, there were few gains in productivity between 1980 and 2000.

Productivity and expansion have together met the challenge of feeding most of the world's population as it has increased to 6.5 billion, while the number who remain hungry, 800 million, has not changed. The challenge, obviously, is to continue to make innovations that will improve productivity and eradicate starvation, without the further expansion of agriculture. For the world's inhabitants who are adequately fed, food in real terms is cheaper and safer to consume than at any time in history. In the affluent world, international trade and transport equipment powered by unrealistically cheap fossil fuels make it possible to buy a larger range of foods than ever before. In more and more countries, the dual challenges of feeding the hungry and providing choice for the affluent will be undertaken within the paradigms of sustainability and connectivity, the subjects of later chapters.

4

PRODUCTIVITY TO EXCESS

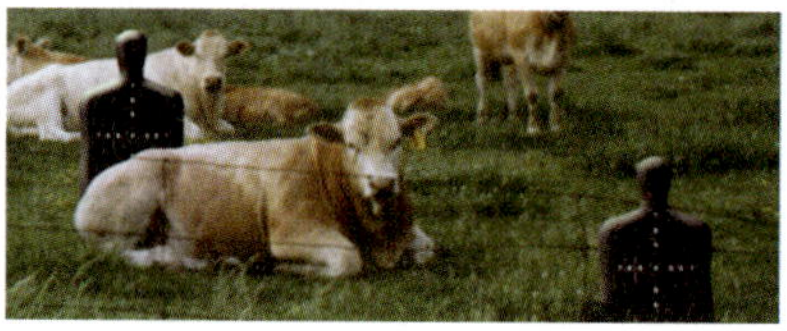

Otis Dozier's lithograph *Grasshopper and Farmer* (Fig. 45) is a powerful comment on the devastating effect of the Depression and the Dust Bowl years on agriculture before World War II. After the war, the paradigm of productivity – the dominant way of thinking among farmers, agricultural scientists and governments making land policies – gained momentum, encouraged by a belief that farm profitability could be improved by producing more food, more efficiently; by concerns about food security (the ability to produce enough food within national borders to avoid dependence on imports); and by urban consumers' delight in shopping for ever-cheaper food in newfangled supermarkets. With few exceptions, societies were supportive – increasing agricultural output was consistent with the idea of "development" and the desire to increase gross national product – or silent. In the 1950s to 1970s, technology was evolving, but non-threatening.

However, while productivity created increasingly inexpensive and safe food, it did not yield the benefits that farmers had hoped for. Farm income in real terms continued to decline: food could be produced ever more cheaply, but free competition among thousands of farmers meant that the cost savings were captured mostly by con-

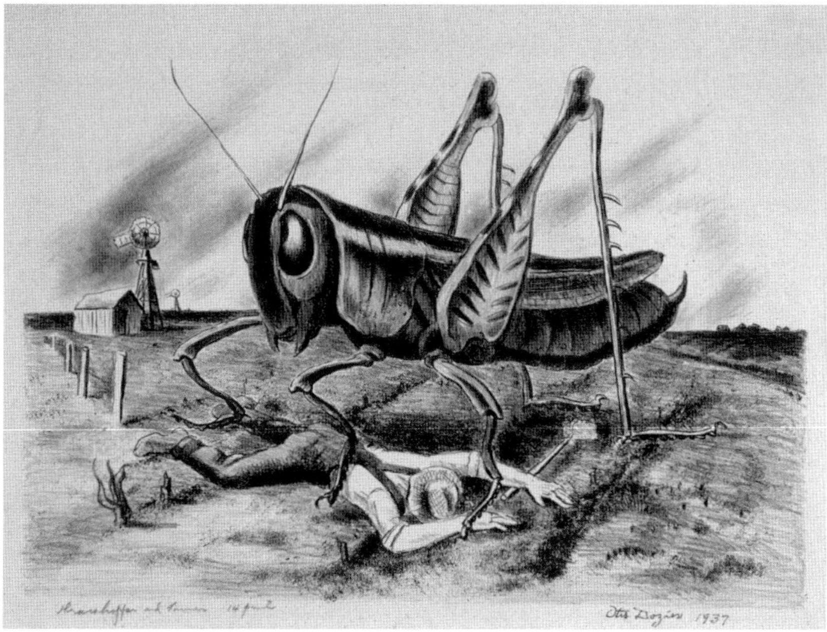

Figure 45
Otis Dozier, *Grasshopper and Farmer*, 1937 (lithograph).

sumers, with some benefit accruing to wholesalers and retailers and none to the primary producers. Even as early as the 1960s, farmers were being seen not so much as "good" custodians of the land, but as "poor."

The financial failure of productivity – its creation of poor, not good or rich, farmers – can be visualized this way. Imagine a graph of gross farm receipts, costs, and the difference between them: the farm family's net income. Since World War II, receipts have increased linearly and steeply. That side of the equation looks terrific. However, for Iowa, Ontario, Argentina or any place for which data have been assembled since World War II, costs have also increased linearly, parallel with receipts. This is because greater inputs have been put into conventional agricultural production systems, and the cost per unit of inputs has continued to rise. Consequently, the third line of our mental graph – net income – has stayed absolutely flat for sixty years. This means that, while some of those farms in Iowa,

Ontario, or Argentina have become larger, more efficient, and more profitable, the "average" farmer has become poorer relative to urban society.

Faith in the paradigm of productivity has made most farmers not only poorer, but also more exposed to risk. Visualize the graph again. In 1950, for example, receipts, costs, and net income were relatively close together. If a crop failed in one year – and bear in mind that the farmer still has to pay the costs whether the crop booms or busts – these costs might have been equivalent to what ordinarily would have been a year's net income. With an above-average harvest the next season, the farmer might have paid off the debt. By 2000, however, costs and receipts might have become six and seven times the farmer's net income, and so a failure would mean a debt equivalent to six years' average income. Only many successive good harvests would save the farmer from bankruptcy.

The preoccupation of farmers and scientists with productivity, together with the "poorness" of farmers, fostered a certain blindness to the negative effects of intensive agriculture. The widespread and often unthinking application of pesticides was the first instance of the paradigm of productivity leading agriculture, and society, into a dark corner. Certainly in the public's mind, as in that of the farmer, pesticides were a godsend, given the experience of the locust infestations of the Dust Bowl years. Dozier's lithograph is an unforgettable image of the farmer crushed not only economically but also spiritually by hard luck and misguided practices. Nor was this to be the only period in which farmers were beset by this kind of double misfortune. In our own time, rapacious business practices driven by money rather than ethics have created unreasonable risks and external impacts. Foot and mouth disease (FMD) and bovine spongiform encephalitis (BSE), popularly known as mad-cow disease, are two demonstrations of the risks associated with modern, highly interconnected agriculture. Both are associated in the public mind with mass production and high-input agriculture, but in fact foot and mouth disease has occurred sporadically throughout history in livestock living in extensive, relatively natural grazing lands, and mad cow disease arises through the feeding, not the housing or general welfare, of cattle. However, both diseases have spread extensively as a result of a few people doing unethical things and the highly complex food chains that were created in the latter years of agriculture's productivity paradigm. Thus, while this paradigm has brought immense benefits, there has yet to be a balanced accounting: Has productivity led to food security in the West, or to "productivity to excess"?

We can think of productivity as being "excessive" when it has a negative impact on, or distorts, the greater physical or social environment, including human health. In agriculture as in any economic activity, any impact on others outside that activity is termed an "externality." An externality affects, whether positively or negatively, the welfare of or opportunities available to an individual or group without direct payment or compensation. As Pretty and colleagues[1] noted in 2000: "The types of externalities encountered in the agricultural sector have five features: (1) their costs are often neglected; (2) they often occur with a time lag; (3) they often damage groups whose interests are not represented; (4) the identity of the producer of the externality is not always known; and (5) they result in suboptimal economic and policy solutions." Interestingly, there are very few scientific studies of the full impact of agriculture, even when, as with the building of high numbers of hog farms, it seems clear that there are large negative externalities. Pretty and colleagues' study of the economic impacts of agriculture in the United Kingdom can therefore serve as a model for the impacts of intensive, productivity-driven Western agriculture more generally.

First, by way of context, between the years 1950 and 2000 yields of wheat, barley, potatoes, and sugar tripled while milk yields per cow doubled in the United Kingdom. This had consequences for the environment, such as rising nitrates in groundwater and rivers; this was occurring in much of Western Europe, as will be described in later chapters. Positive externalities also arose from the intensification of agriculture, such as the maintenance of trees for aesthetic reasons and carbon sequestration, and the careful use of landscape with the goal of storm protection and flood control.

Externalities may be grouped into categories such as things that damage (or benefit) aesthetics, biodiversity, human health, and the environment (Figs. 46, 47). Pesticides pollute groundwater, streams

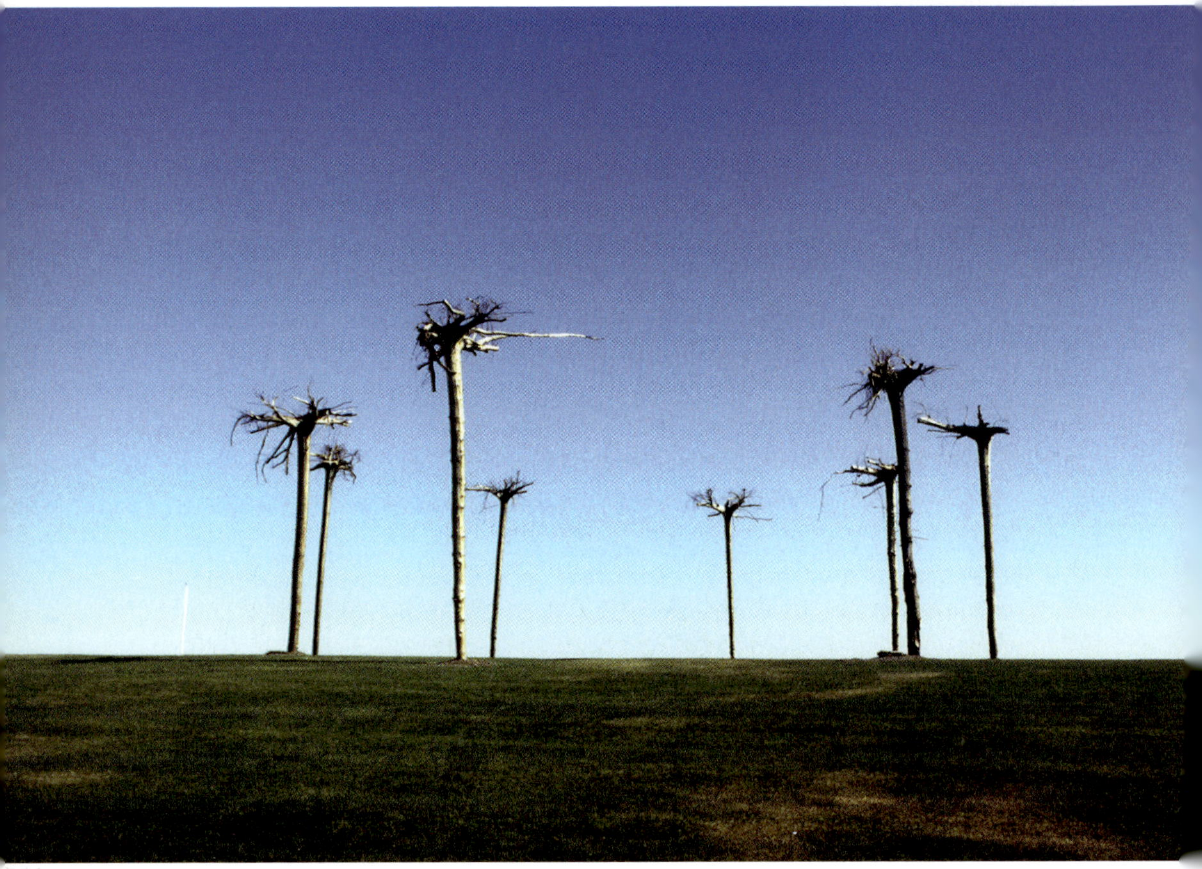

Figure 46
Reinhard Reitzenstein, *No Title*, 1987 (trees inverted and peeled). Reitzenstein's sculpture of eight unearthed trees speaks of the jarring effects of environmental pollution. This direct statement is a powerful warning: the root ends of the trees suggest gaunt human figures.[2]

and rivers. So do nutrients such as phosphorus in what is referred to as point-source pollution, especially from hog barns, and nitrogen which is a major element of all manure. Nitrates from excessive use of fertilizers on wheat and corn fields, and from animal manures, may cause eutrophication (the over-concentration of nutrients) of ponds and waterways, which in turn has negative impacts, such as loss of biodiversity and of the aesthetic value of recreational areas. Farm wastes, such as slurry from settling ponds on live-

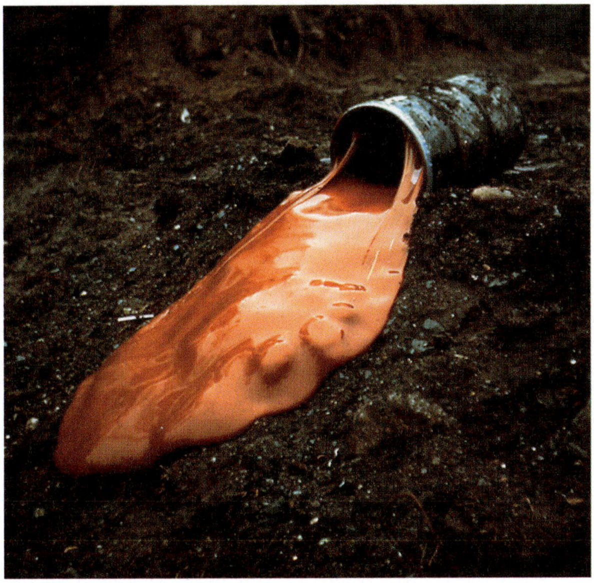

Figure 47
Robert Smithson, *Glue Pour, Vancouver, BC*, 1969 (aluminum drum, glue). Smithson's sculpture, an early response to public concerns about the chemical contamination of soil and water, reminds the viewer of rotting effluent pipes seen along river banks near old industrial sites. One wonders what potential for harm these noxious substances contain.

stock operations and silage effluent, frequently cause concern. In each of 1996 and 1997 there were about 31,000 reports of water pollution in England and Wales from all sources, and throughout the 1990s some 2,600 cases of agricultural pollution were recorded annually. Micro-organisms from industrial agriculture also enter waterways and, like nutrients, give rise to direct costs associated with ill-health, government legislation and monitoring and, most particularly, the need to deliver clean drinking water to cities.

Although public awareness of these agricultural "bads" has grown, mostly since the 1950s, negative externalities still occur. For example, in 2000, manure from livestock feedlots caused *Escherichia coli* to enter the water supply of the town of Walkerton, Ontario; six people died (four directly from *E. coli* infection, two from complications of *E. coli*). Five years later, anyone driving past corn and soybean fields in the same region during the hot summer of 2005 would see one reservoir after another posted with signs warning: "Closed: water unsafe for swimming/drinking." These costs, of course, are borne not only by country dwellers but by urban taxpayers and consumers as well.

Agriculture is our primary vehicle for keeping the air clean, through growing crops. It is also a major source of pollutants in the form of nitrous oxide lost to the air from crop fertilizers; methane (cows and horses are ruminants and burp significant quantities of this greenhouse gas into the atmosphere); ammonia from urine and feces; and carbon dioxide. The biggest store of carbon or organic matter is in the soil. Soils release carbon as carbon dioxide when organic matter is broken up by ploughing, as when pasture is converted into cropland. The total amounts of carbon are prodigious; estimates for the United Kingdom place this at about two hundred tonnes of carbon per hectare under permanent pasture and crops. If farming practices wisely husband or increase this carbon, the amount of carbon dioxide released into the atmosphere will be reduced, slowing down the negative consequences of climate change.

Agriculture has a range of other externalities that could themselves be the subject of books, such as its impact on biodiversity and on landscape aesthetics. Considering the effect of large-scale agriculture on larks, hedgerows, and drystone walls leads to the question: How does society value agricultural aesthetics? There were once roughly 120,000 kilometres of drystone walls around fields in the United Kingdom, of which about 7,000 kilometres have been lost altogether. About 45 per cent have become derelict over the past forty years, and almost all need restoration. The Countryside Stewardship Scheme pays farmers or others to restore these "externalities," which have been costed at about £24 thousand per year.[3] By contrast, we do not yet have mechanisms to put costs on the ongoing loss of biodiversity.

Although industrial agriculture has created uniform, low-cost food for the consumer, it has also aggregated, or focused, supply systems such that they are more easily monitored and their quality better controlled. The effect is safer food than ever before. This safety is increasingly underpinned by legislation and audits, as discussed in chapter 6. However, there is another side to a productivity-driven Western agriculture, namely the externalities relating to human health. Pretty and colleagues identify just three elements to this, although they are more plausibly examples rather than a complete list: bacterial and viral outbreaks in food; antibiotic resistance; and bovine spongiform encephalopathy.

First, there are half a dozen common organisms that may cause widespread food poisoning (Fig. 48). As people have become more observant and litigious, the reported incidence of food poisoning

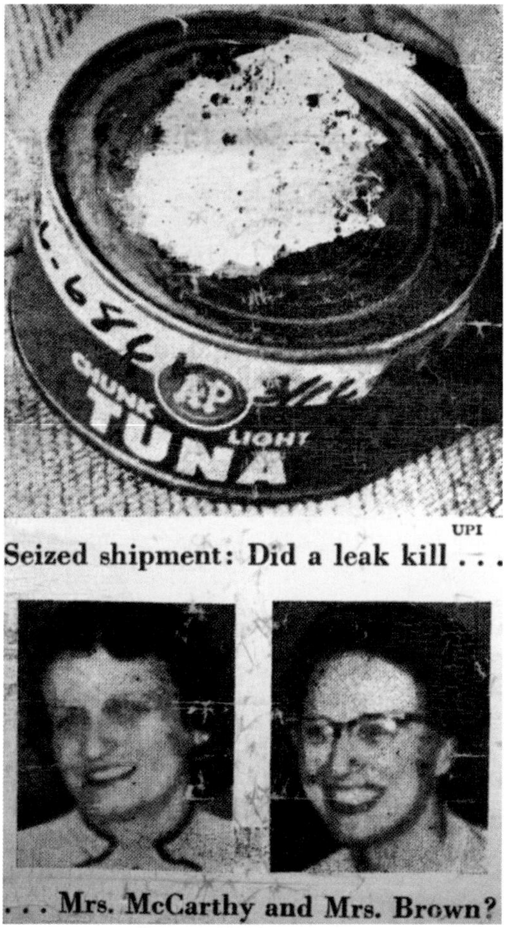

Figure 48
Andy Warhol, *Tuna Fish Disaster*, 1963 (silkscreen ink and silver paint on linen). Warhol, perhaps best known for his depictions of Campbell's soup cans and Brillo boxes, comments on food poisoning in this dramatic work. Artworks like this painted serigraph, as well as extensive television coverage of individual cases, raised public awareness of this important issue.

has risen sharply in recent years, reaching 94,000 cases in 1997 in the United Kingdom alone. However, this represents the tip of the iceberg: many cases go unreported, and people sometimes die from food poisoning without it being diagnosed. It may be that only one case in thirty is reported, although it should be borne in mind that farming itself probably only causes 20 to 25 per cent of the total, the others being the result of unsafe food handling in the home, restaurant or market.

Pretty and colleagues' second example of an aspect of industrial agriculture that affects human health is the veterinary use of antibiotics, that may lead to resistance in farm animals and humans. Over a thousand tonnes of antibiotics are used in the United Kingdom each year, of which one-third are used for farm animals. They are used when animals are sick and – the majority of cases – to stimulate growth or prevent illness in large-scale, intensive production units. Although one might question the benefit of this practice to the consumer who eats the food, there are also concerns that agriculturally administered drugs will ultimately lead to antibiotic resistance in the human population, making our own bacterial illnesses more difficult to treat (Fig. 49).

The third example that Pretty and his colleagues consider is that of bovine spongiform encephalopathy (BSE), dubbed "mad cow disease" in view of its late-stage symptoms. The staggering costs of BSE in the United Kingdom have been conservatively estimated by Pretty and colleagues at £600 million in each of 1996 and 1997,

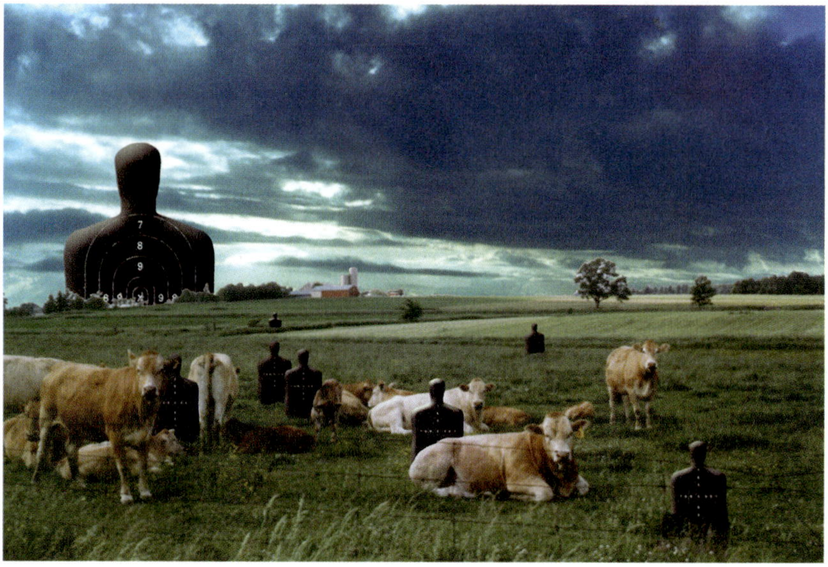

Figure 49
Robert Howson, *Boo Moo*, 2002 (digital print). Howson takes aim at the vulnerability of the cattle farm as we know it. This image suggests that not only are cows fed an unnatural brew of antibiotics and animal protein, but that this practice can have the consequence of making them sick.

before the outbreak peaked. Because the outbreak of BSE and its management by government is an instructive example of how the pursuit of productivity can go badly awry, we will return to it in more detail shortly.

Considered all together, and allowing for some stupendous approximations and assumptions, the negative externalities associated with agriculture in the United Kingdom amount to more than £2300 million each year. Of these, the big costs relate to gaseous emissions (about half the total), and the impact on human health through BSE and food poisoning. The total is equivalent to £208 per hectare of farmed land, or 13 per cent of gross farm income. Some of the costs of these externalities can be attributed directly to inputs. For example, pesticide externalities related to cleaning up drinking water amount to £8.60 per kilogram of active ingredient sold. This suggests a fairly simple policy option: namely, adding this cost to the price of the pesticide so that (rather than a tax on all people) agriculture, and later the person who consumes the food, pays directly for its production. Other externalities relating to industrial agriculture are less easy to predict or quantify: as livestock production units become larger, so do the risks to human health from, for example, poisoning of milk from a large-scale regionally integrated dairy chain, or an emerging disease akin to BSE.

One disease that Pretty and colleagues' study did not consider is foot and mouth disease (FMD), which recurs sporadically in many countries and so is not necessarily associated with highly productive, complex, or industrial Western agriculture. FMD is a severe, highly communicable, viral disease named for the blisters it produces on the tongue and around the mouth or on the feet (in the softer tissue between the hooves) of cattle, pigs, sheep, goats, and deer. It does not necessarily kill the animal, but causes excessive salivation, lameness, and debilitation: cows have lower milk yields and are more prone to diseases such as mastitis; spontaneous abortions increase; calves are less likely to survive; and growth is stunted.

The FMD virus, which survives in lymph nodes and bone marrow, has at least seven separate types and many subtypes. Immunity to one type does not protect an animal against other types. Unfortunately, with the right temperature and pH conditions the virus can persist in contaminated fodder and the environment for up to a month, which allows it to spread by wind and through the transportation of animals or feed. The good news is that it does not affect humans.

The United States has been free of FMD since 1929, when the last of nine outbreaks was eradicated. In Europe, and especially in Italy, outbreaks occurred regularly until the late 1980s. Periodically, cattle were slaughtered, and vaccination, carried out annually until 1991, almost eradicated the disease. There has never been a reported case in Australia, whereas Argentina, with a similar climate and grazing lands, is beset with the problem: FMD was reported fairly regularly until the "last case" was reported in 1994, and this caused a quarantining of exports until 1997. With no further outbreaks, that country was declared FMD-free in May 2000. Only months later, FMD antibodies were detected in Argentinian cattle, and by February of the following year infected animals were found. Once again, exports were quarantined. FMD was subsequently found in pigs in 2003. Today, the country is split with respect to FMD: countries that are free of the disease, such as Australia and the United States, will not import meat or feed from north of latitude 42°, where FMD is present. In the south, Argentina remains FMD-free. To take another example, India has endemic FMD; it has been estimated that the disease causes a 6.5 per cent loss of total milk production (about 18 million rupees per annum) and an equal cost arising from lost draft power, animal deaths, and the cost of treatment.

FMD was brought to the world's attention in 2001, when it was again found in England (Fig 50). Twenty years of secure status were swept away, and the public was treated to graphic images of how quickly a contagious disease can spread in a complex agricultural system in which animals and feed are moved all over a country, albeit a relatively small one, every day. The English FMD virus, according to a government enquiry[4] was most probably introduced from "the Far East," the source of an earlier outbreak in South Africa. It is likely that meat was illegally imported from Asia and perhaps used in restaurants before the scraps were discarded and fed to pigs. The infected meat did not harm the human consumers, but infected the pigs. By the time the pigs carrying FMD were discovered at an abat-

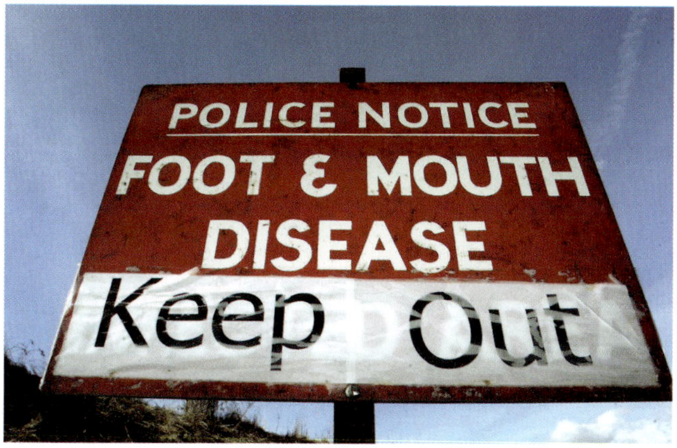

Figure 50
Police warning to keep out at Burnside Farm, Northumberland, where the UK foot-and-mouth crisis is thought to have originated. (Associated Press, 15 March 2001).

toir in Northumberland, more than one lot of infected pigs had been processed and the virus had spread from the abattoir through "mechanical and personnel transmission." Meanwhile, wind blew the virus from the infected pig farm to a neighbouring property, infecting sheep that were subsequently marketed and dispersed across England, Wales, and southern Scotland. A total of over two thousand cases of FMD were created between 20 February and 30 September 2001. Initially, the ministry was confident that, by means of exclusion zones and killing animals, the disease would be contained quickly. However, officials underestimated the extent of animal trans-shipments by half; as contemporary writers expressed it: "Nobody took account of the extent to which dodgy farmers moved sheep around to claim quota payments" and "[t]he disease was spread to Scotland, Northern Ireland and the Republic of Ireland by a single dealer looking to gain £10 per head premium plus a 4.5% VAT [value-added tax] rebate by illegally passing off English lamb bought at Carlisle as Irish lamb."[5]

In one or two days, complex animal trans-shipments and delays of some hours caused the outbreak to assume horrendous proportions. Two thousand animals were found with FMD; 10 million

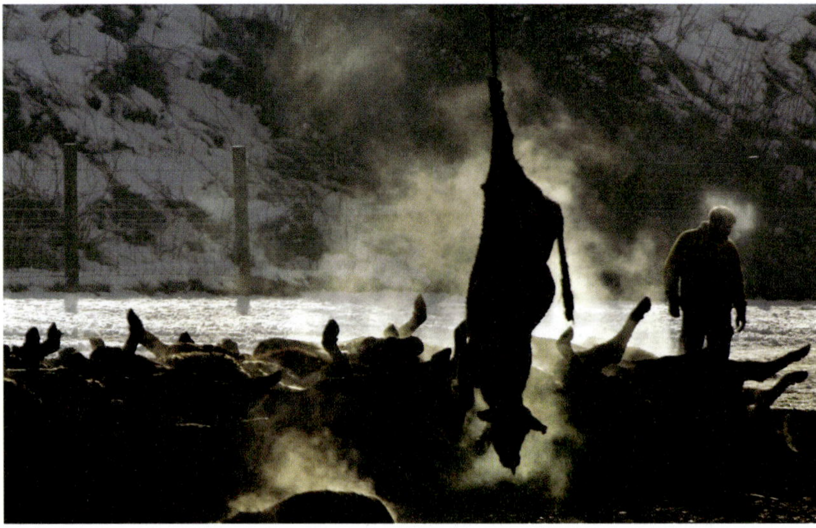

Figure 51
Disposal of cattle at Lockerbie, Scotland (Associated Press, 3 March 2001).

cattle, sheep and pigs were slaughtered. Transport of animals and people ground to a halt. The scale was quite unlike the earlier, sporadic outbreaks throughout Europe, in which (on average) about twenty-nine holdings would be affected before the outbreak was halted (Fig. 51).

The slaughtering and disposal of carcasses was a mammoth undertaking conducted under the eye of the international media. After an unconscionable delay, the army was mobilized: Brigadier Alex Birtwistle, charged with the clean-up operation, estimated that about a hundred thousand carcasses had been waiting for disposal by the time the army was called: "It was a hell of a mess by the time we were brought in. There were rotting heaps of carcasses that had been outside people's houses for about three weeks."[6] Cows had not been separated by age: those over five years old needed to be burned, whereas those under five years old could be buried. Animals were so badly decomposed they could not be accounted for. As Birtwistle described it, "Their ears were so rotten their ear tags had fallen out so we couldn't identify which was which"[7] (Fig. 52).

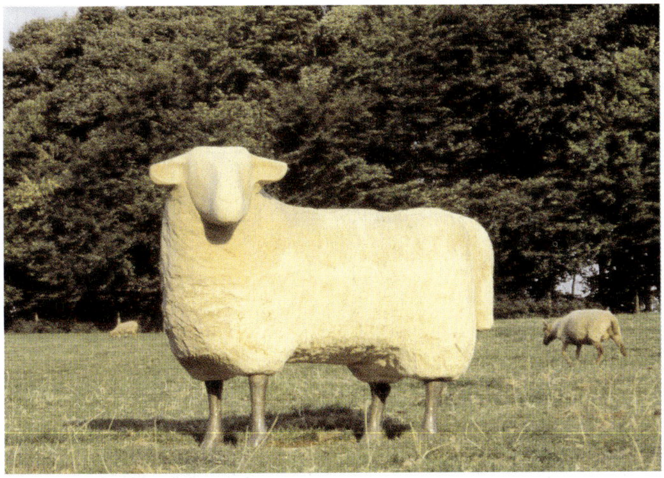

Figure 52. Diane Maclean, *Wether*, 2001 (Clashach sandstone and bronze). Scottish artist Diane Maclean responds to the horrors of the foot and mouth disease epidemic with this sculpture: "*Wether* is my memorial to the hundreds of thousands of sheep who were slaughtered (mostly needlessly). Made of sandstone and bronze, it stands in a sheep field. It is an archetypal sheep. The black dots on the head are flies."

The fragility of modern agriculture was exposed through nightly news broadcasts. That fragility was costly: one estimate is that the 2001 FMD outbreak in the UK cost about £3.38 billion.[8] Agricultural production was, not surprisingly, affected, but only by about 30 per cent – not much greater than the cost of additional staff in the government agencies and armed services that responded to the outbreak. The largest agricultural costs were in fact related to the longer-term reduction in production and earnings from the livestock sector and the short-term compensation payments made to farmers as their stocks were slaughtered. However, the direct costs to agriculture was small in relation to the indirect impact on the larger, tourism sector: here, the loss was about £7.73 billion. Hotels, catering and pubs, railways and road transport, retailers and others were all hit, as all country walking paths and most inland waterways were closed to the public. Many tourist attractions, such as Stonehenge, zoos, and grand houses, were out of bounds. As the impact of FMD on agricultural production lasted several years, so did the impact on tourism, which probably did not return to normal levels until 2004. FMD broke out again in England in 2007; it is too early to predict its direct and indirect costs.

Bovine spongiform encephalopathy (BSE) is a disease of a different complexion. It is one of the rare diseases that can be transmitted from one generation to the next, and from livestock to humans. BSE could, and perhaps did, arise through a naturally, randomly occurring genetic transformation; possibly it arose in a flock of sheep in an idyllic, pastoral situation not unlike those described in chapter 1. However, it became epidemic among cattle, and infected and killed people, because of the complex conditions of contemporary, industrially oriented agriculture. As the report of the Horn review, an inquiry into the origin of the epidemic, stated in 2001: "The BSE epidemic has already caused untold human suffering; it has severely damaged the farming and associated industries in the UK and also put these industries in other countries at risk. It is important for future generations that we understand and learn from the lessons of the recent past."[9]

Cattle are fed supplements so that they grow faster or produce more milk, or simply because they are housed in such concentrations, or live in climates that have seasons with poor feed supply, and so require extra "hand feeding." The BSE story starts with one innocuous-sounding supplement: bone meal. For centuries, cattle have been fed meat and bone from dead or slaughtered animals to supplement their intake of protein and of minerals such as phosphorus. In the nineteenth century, this practice led Gilbert to his study of the more widespread use of phosphorus and to the discovery of superphosphate. Bone meal was, at least until the 1980s, increasingly fed to cows as a high protein supplement when they started milking (Fig. 53). Hence, they were commonly exposed to it at two or two and a half years of age. However, in the United Kingdom (and in Australia), it was also common to feed bone meal to calves, starting at about two weeks of age, when they are separated from their mothers in commercial dairy farms.

It is thought that a novel, highly infective spongiform encephalitis originated in the early 1970s through a mutation from the naturally occurring (and not harmful) prion protein to create a new shape of prion protein in cattle and likely, sheep. Whether this was triggered by natural random mutation or by an environmental agent such as a toxic chemical will never be known. The normal prion

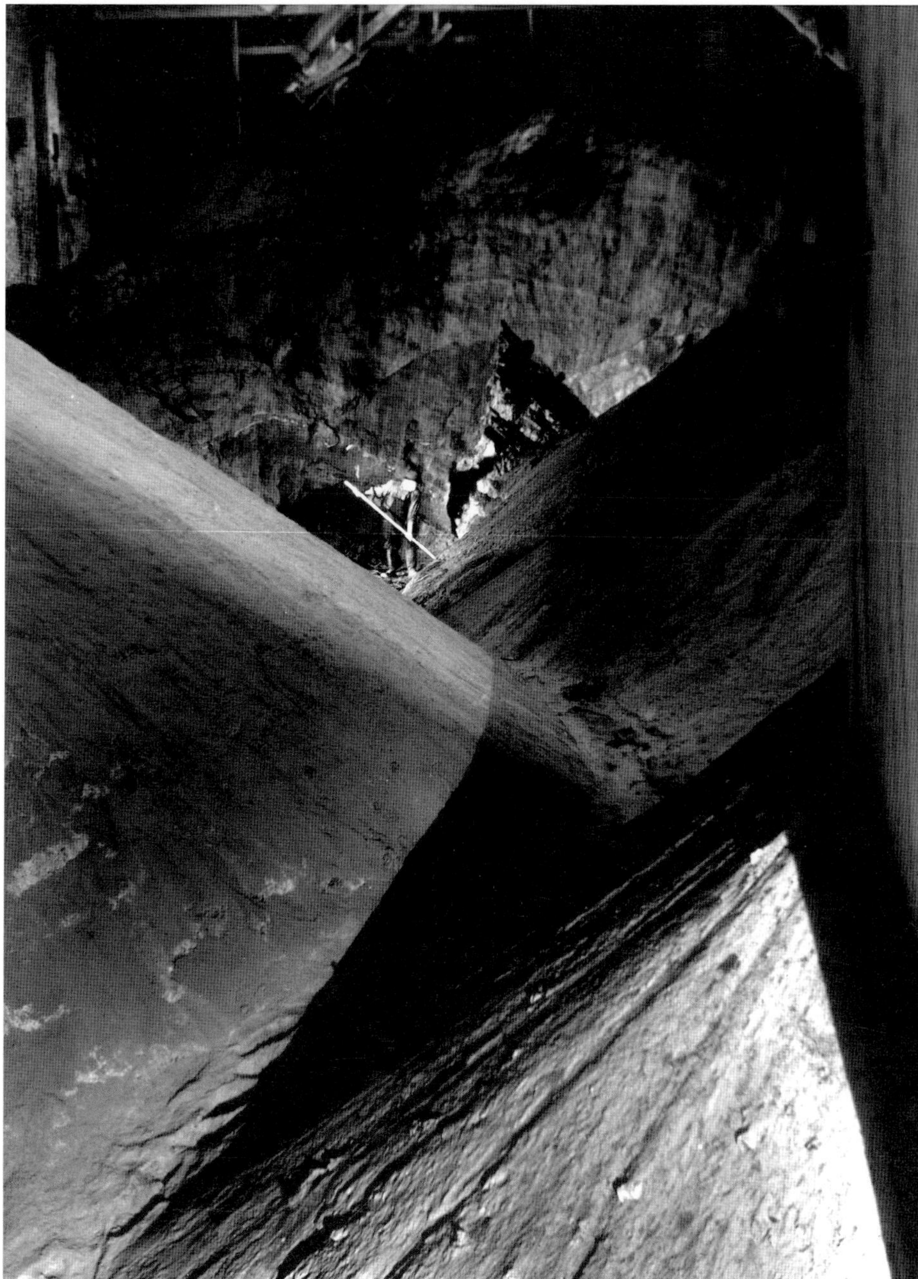

Figure 53
Margaret Bourke-White, *Render in Chicago*, 1930 (silver gelatin print). Famed American photographer Margaret Bourke-White documented the hog rendering in a photo-essay on Swift and Company for *Fortune* magazine. Her photograph shows a single worker among enormous mounds that are "the final remains of the pigs ground to an acrid powder that is later mixed with meal and sold as fodder."[10]

protein is a membrane protein that occurs commonly in many tissues, including the central nervous system and brain of healthy animals and humans. Variants of the normal prion protein are common in sheep in some countries, causing a disease called scrapie, which is not lethal. It is not known whether the mutation arose from the scrapie prion, which would account for its occurrence in the United Kingdom, where scrapie is endemic; if this is the case, however, the presence of the prion gene in Australia is not easily explained, because that country is scrapie-free. Or it may simply be a very rare, random mutation unconnected with scrapie or sheep.

David Waltner-Toews, a professor at the Ontario Veterinary College and a poet, has written about kuru, one of several known transmissible spongiform encephalopathies, which behaves much like VCJD:

> It begins with eating the brains
> of those whom we admire,
> with women and children first.
> It ends with Kuru, a spongy Jacuzzi
> of laid back prions, engulfing the brain.
> It ends with a young woman
> throwing herself into the fire.[11]

In any event, it is most likely that the new, virulent prion infected British cattle and spread through the carcasses of infected livestock that were ground up into bone meal and fed to calves. It is likely that BSE was first, and most severely, found in the United Kingdom because it was there that makers of high-protein feeds began in the early 1970s to include bone meal in the ration fed to calves. Elsewhere, in continental Europe and North America, bone meal was extensively used as feed for milking cows but rarely for calves.

There were likely several cycles of BSE in the southwest of England in the 1970s and 1980s that went undetected: affected cattle were probably considered to be aged or to be suffering from severe magnesium deficiency, an illness that shows similar symptoms including a lack of coordination or "madness." These cattle would have been quickly culled, their infected meat introduced into the food chain, and their bones distributed through the stock-feed system. The effect was disastrous: as little as one gram of infected bone meal can, if eaten, eventually cause BSE and death.

The normal membrane prion does not cause harm. The variant or virulent prion does. Nastily, it also spreads: contact between normal and abnormal prions induces the normal to convert to the abnormal form, which is resistant to degradation normally associated with ageing. Thus, once they have infected an animal, the abnormal proteins build up, spreading along the nerves to reach the brain, where it spreads further. Because these nerves cannot be repaired or replaced, they become mushy or spongy: the disease incubates for two to eight years as it spreads, and then symptoms of "madness" appear. Death occurs in cattle from two weeks to six months after the symptoms start.

"Mad" cattle with spongy brains were identified in the 1980s. The first cow to be recognized as having a peculiar new disease died in February 1985. The number of cases peaked in 1992 at 36,680, falling to fewer than 1,500 cases annually since 2000. The United Kingdom responded by prohibiting the feeding of bone meal to livestock and consumption of any part of slaughtered infected cattle. More recently, all brains and spinal cords, whether infected or healthy, are incinerated to avoid the chance that they may be recycled or eaten.

Incredibly, bone meal was exported. A domestic ban on bone meal imposed in 1988 precipitated a boom in exports until 1992, when abattoirs were required to remove and incinerate spinal cord materials, and bone meal was again considered safe. But, through these exports, BSE has arisen elsewhere in Europe and, in 2001, in Japan. The Japanese cow, the first identified outside Europe, was most likely fed contaminated bone meal imported from the United Kingdom. Only 194 tonnes were imported directly in 1990 and 1991, but a much larger amount was imported via Indonesia.

Meanwhile, residents of the United Kingdom were eating meat that, at least in very rare events, contained the BSE prion. In 1996, a new variant of a naturally occurring but very rare disease, Creutzfeldt–Jakob disease (CJD), was identified. This variant, vCJD, is caused by infection from the prions that caused BSE in their former, bovine hosts. To date, 135 people have died worldwide from VCJD, and more deaths are expected. Most recent models put an upper estimate of seven thousand deaths over the next thirty years, in the absence of further international spread of BSE.

Estimates of the cost of the BSE epidemic have not to our knowledge been published, but the cost of surveillance schemes and health care, and the loss of consumer confidence in beef, even temporarily through the 1990s, far outweigh the costs of slaughtering thousands of cattle.

In May 2003, a single mad cow was discovered in northern Alberta; almost three thousand animals were slaughtered, none of which had the disease. The real cost is still unfolding as the Canadian beef industry restructures: about one million tonnes of Canadian beef were being exported to the United States each year, and this stopped within hours of the discovery of BSE. In December 2003, a single Holstein cow tested positive in Washington State. This animal came from Canada as part of a herd that was sold off several years previously. Although these cases are both isolated and rare, government responsiveness was encouraging. Within seven days of the preliminary diagnosis, all the calves of the infected cow had been traced to various farms, whose herds were quarantined. The seventy-three cows that accompanied the Holstein now carrying BSE from the original herd in Canada were traced, despite the passage of years, and meat products from the abattoir that killed the infected cow, now distributed to forty-two locations, were recalled. Meanwhile, in Europe, long-term, uneconomic storage of unusable meat and bone meal is creating pressure to relax the stringent prohibition against feeding animal by-products to animals, such that animal wastes may be "cross-fed" from one species to another (but avoiding ruminants). Although the BSE epidemic was not a foreseen consequence of industrial agriculture, it is by virtue of the very complexity of industrial agriculture, and its high level of organization, that we have the capacity to oversee our food chains and to ensure that, through rapid response, we minimize the impact of "the next BSE"– whatever that might be.

Pretty and colleague's study of agriculture in the United Kingdom attempted to quantify its value, and its cost, to society as a whole. To this, one could add the cost to other societies of Western agriculture's production of excessive amounts of food. In the absence of trade barriers, low-income countries might find it easier to produce food, either for themselves (in the absence of cheap, dumped food from high-income countries) or for export to more affluent countries (in the absence of trade barriers). The topic of European subsidization of agriculture to achieve both food security and a populated countryside, is touched on in chapter 5.

The impact of productivity-driven agriculture is most noticeable in industrial livestock operations. Large-scale chicken farms and hog farms are like factories – which is not intended as a disparaging remark, for factories abound in contemporary society. The sizable agricultural "factories" that are filled with animals produce large amounts of animal feces and, particularly in the case of hogs, noxious odours, and are sited out of view. But to drive down side roads near Brandon, Manitoba, or Griffith, Australia, or a hundred other towns, is to pass rows of long, galvanized iron sheds containing caged chickens or hogs, or to pass through landscapes literally full of cattle in open pens. From the air, multi-toothed galvanized roofs are visible, sometimes tied to their landscapes through lakes of manure in settling ponds. Writer Annie Proulx describes "scores of anonymous, low, grey buildings with enormous fans at their ends set back from the road and surrounded by chain-link fence. From the air these guarded hog farms resembled strange grand pianos with six or ten white keys, the trapezoid shape of the body the effluent lagoon in the rear."[12]

Inside, the chicken broiler or egg barn presents rows and rows of fowl in cages, three to five per cage, to grow or produce eggs that roll out and along conveyor belts. The cages are stacked one on another, three high, and the dim artificial light never stops. Neither does the food, which today's chickens are five times more efficient at converting into meat than the chickens of thirty years ago. Egg-laying must be approaching its biological limit, with barns averaging 350 eggs per hen in a year. The air, although humid, is clear and without odour.

By contrast, hog barns *do* smell. Although the barns are clean, automated and relatively humane, the feed rations and the genetics of the pigs themselves cause smells. Pigs are noteworthy, or notorious, because they do not utilize much of the phosphorus in their diet: although they are able to absorb almost all of the two most important elements, energy and protein, from feed, they extract only a relatively small, variable amount of phosphorus: 14 to 50 per cent of the phosphorus in grains and grain by-products, and usually less than 30 per cent of the phosphorus in protein supplements such as cotton seed or canola meal. The rest passes through as the hogs' feces to the environment.

Earlier, the problem of low phosphate absorption and excessive excretion was addressed by feeding the pigs meat and bone meal, from which the phosphorus is at least 90 per cent absorbed. Feeding animals meat and bone meal is now banned, at least in Europe, America, and Japan, to decrease the potential spread of serious diseases such as BSE and FMD. However, it remains a temptation for low-income or unscrupulous producers, with the ever-present possibility of disease outbreaks that may cross the species barrier to humans.

As a substitute for bone meal, animals are fed plants such as grains with supplements of phytase, which breaks down the phosphorus in the grain and makes it available to the pigs. However, this is expensive: although it decreases the phosphorus in the feces by 25 to 50 per cent, it costs the individual producer money and is prohibitively priced for small producers in low-income countries.

High-phosphorus feces are smelly and contaminate groundwater and waterways. In the next chapter, we will discuss water pollution in Europe and North America, but it is worth noting here that the problem is not confined to intensive, industrialized agriculture: in Santa Catarina, Brazil, where hog farming involves a mix of industrial barns and small-holders, swine production and the spreading of manure has caused 85 per cent of water sources to be contaminated by fecal organisms. The contaminated water smells, like the air downwind of hog barns.

Industrial agriculture is based on two logics. First, the real price of food to the consumer has fallen throughout the era of productivity. This long-term trend is not about to change and, indeed, industrial agriculture contributes mightily to its continuance (Fig. 54). Second, creating an increase in productivity, so that input costs fall more than output prices, requires an agglomeration of production units. Operating on a large scale is the secret, and the economies of scale are enhanced if, as usually happens, governments support large-scale operations in ways that effectively reduce an operation's input costs: through taxation breaks, by providing easier emission standards, or by not requiring the agricultural operation to bear the full costs of its externalities through, for example, waste management and odour abatement.

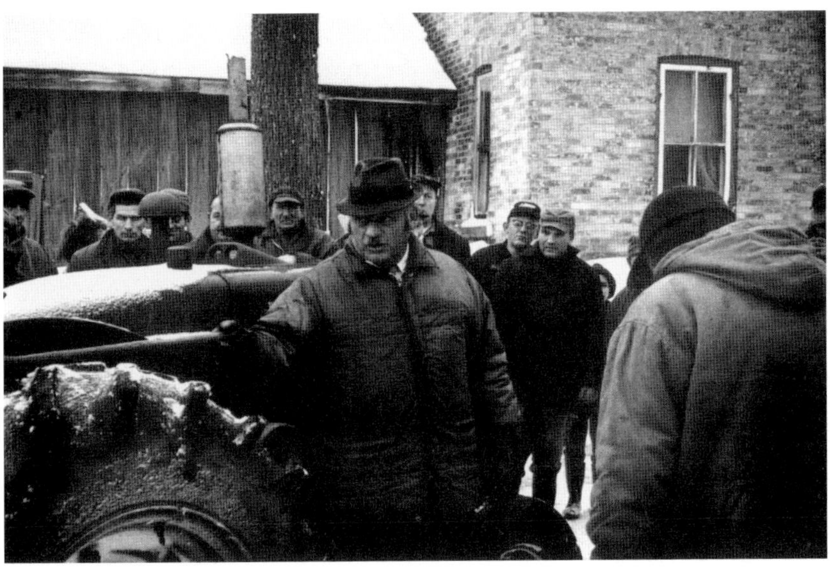

Figure 54
David Nasby, *Max Storey, Auctioneer*, 1972 (silver gelatin print). Some documentary photographers have recorded the social effects of industrial agriculture. This photograph of an Ontario farm auction shows the auctioneer in full swing as he disposes of the goods and chattels of a century-old farm family that can no longer compete. Welfare activists also periodically confront companies and universities about the ethics of producing, or researching, cheap food at the expense of animal rights and welfare and of biodiversity.

Industrial aggregation aims to reduce costs faster than the reduction, in real terms, of output prices. Its extent can be impressive. The 2003 website of Agriculture and Food Manitoba advertised that pork is Manitoba's most valuable industry, producing 6.3 million pigs worth $860 million per year in 2001. This aggregation had been achieved in less than twenty years: the number of pigs per farm has increased from fewer than 200 in 1976 to over 1,500 in 2001, and (apparently better still) 11 per cent of pig farms now produce 82 per cent of the pigs. Numerous studies of the undesirable impacts of industrial agriculture (in this case, hog farming) have concluded that intensive livestock operations:

- decrease local economic growth
- decrease the value of local properties
- decrease quality of life, measured by, for example, the frequency with which residents can open their windows
- increase the occurrence of adverse health effects such as headaches, excessive coughing, and stress
- increase violations in environmental standards
- change the class composition of society through the introduction of transient labour, shifting the balance toward poorer people with limited attachment to a locality
- increase crime rates and social conflict
- decrease participation in community social life and organizations such as voluntary community services
- achieve financially calculated high productivity in part through the under-valuing of external inputs, such as community-supplied infrastructure (for example, roads) and energy
- create a moral hazard or inducement for subsidies or support from governments, leading to higher costs to regional resident taxpayers
- maximize profitability by shifting the costs of waste management to other residents in the region
- take advantage of government policies, for example, taxation rules that allow them to under-state or eliminate some of their costs
- create self-contained operations that "cannot aid or enhance regional economic development"[13]

Hog farms are preferentially located by corporate business in areas that have rural populations of low social status and incomes, and, according to Wing and colleagues, "most hog operations, which use waste pits that can contaminate groundwater, are located in areas with high dependence on well water for (human) drinking."[14] The case against industrial, aggregated agriculture has been summarized with pithy phrases such as:[15] "it is more advantageous for Iowa to have more hog farmers rather than more hogs."

Industrial hog, chicken, and beef farming appears to be the end-game arising from 150 years of striving for productivity. However, before being one-sidedly critical, it is important to consider also the benefits. Meat and egg prices are lower in real terms, and quality controls higher, than ever before. The technical efficiency of production (for example, kilogram of beef per kilogram of feed or hour of labour) are higher than ever before. Consistency of quality – the toughness or otherwise of the steak – is better than ever before. Consumers have a choice and, to date, have voted in the supermarkets: only a small (albeit growing) percentage of consumers in Europe and the New Worlds are prepared to pay substantial premiums for food produced in less productive, but arguably more humane and less polluting, agricultural systems.

Also, while recognizing the "bad" externalities associated with the hog and chicken barns and feedlots of industrial agriculture, it is important to acknowledge the substantial ongoing research that aims to reduce or eliminate problems with, for example, odour abatement and the disposal of effluent. The application of these research results, when they come, will address some of the dark corners we referred to at the beginning of this chapter. But it is also clear that change necessary to reduce the "bad" externalities associated with productivity-driven agriculture is inherently difficult. Productivity and its end-game, industrial agriculture, presents us with the "problem of the commons:" even if individual farmers had the means to contribute to addressing the externalities that they create, productivity-created problems, such as the fouling of groundwater, soil erosion, or salinity, affect much wider areas than an individual farm. And, because we, as a society, do not yet place enough economic value on the quality of our natural resources, there is not an adequate, immediate incentive for individuals to take action. Some might argue that this is of little concern, as consumers ultimately dictate what they will purchase and how it is produced. There is some evidence of this, such as the rise in demand in affluent societies for organically produced food, or food produced using ethical labour practices. However, we should not be complacent. As John Kenneth Galbraith, the economist trained in his youth as an agriculturist, railed: "Belief in a market economy in which the consumer is sovereign is one of our most pervasive forms of fraud."[16]

The problem of the commons, and the fraud of consumer sovereignty, are currently addressed by an ad hoc mixture of government regulations and collective, voluntary action, as through Landcare groups, to clean up externalities or keep them in check. This is a long-term and collective solution, but one that is difficult to implement in a way that gives the urban public a feeling of security with respect to the fear that negative externalities and commercial interest might some day cause a disaster for the consuming public. Ad hoc, long-term and collective actions contrast with the paradigm of productivity, whose rewards are felt in the short term and are enjoyed by individuals.

5

SUSTAINABILITY

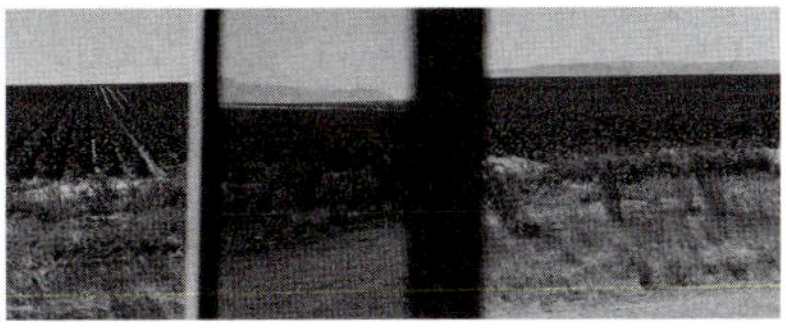

Nowadays, more people are fed, more safely, than at any other time in history. To achieve this, some agricultural lands, such as the rice paddies of Java, have been cropped continuously for more than a thousand years, without loss. But some agriculturally based civilizations have blown away with the wind. The Biblical Garden of Eden, representative of the partly irrigated "fertile crescent" of the Euphrates, blossomed and then returned to semi-desert a millennium ago. We repeat history. Moving land-clearing and agriculture into unsuited areas of Australia, where the geology reflects ancient sediments laid down by salty seas, has caused dry-land salinity on a continental scale. In Australia, fifteen million hectares of land, sixty-seven thousand kilometres of road, and over two hundred towns might as a consequence be abandoned within the next fifty years.[1] The paradigm of increasing productivity has resulted in polluted waterways in Europe, North America and through Asia and crop-induced desertification in sub-Saharan Africa. It has created industrial livestock production units that appear to be energy-inefficient and that in some cases cause harm to their human neighbours. Where is the balance?

Generations of migration from agricultural regions to towns and cities, most particularly during periods of dramatic population growth in the twentieth century, together with the scaling-up and specialization of agriculture so that fewer and fewer farm families are needed or supported in rural areas, creates an ever-widening gap between agriculturally based regions and our cities. Where are the links to bridge the widening gap?

Throughout history there have been many farmers who were concerned about their land resources and managed their land and their crop cycles to preserve or enhance fertility – to "sustain" their resource, as well as their families. Until about the 1940s, this personal preoccupation with sustainability had little voice and no place in the public conscience, or in public policy. Then, through the 1940s and under the pressure of key events in the 1970s, there grew a widespread anxiety about the future until, in the 1980s, governments adopted policies of "sustainable development" or "sustainability." As with any paradigm, our understanding of these concepts has evolved. But it has done so quickly, so that early in the twenty-first century the paradigm of sustainability provides an alternative, or at least a counterweight, to the paradigm of productivity.

In the midst of war, four church leaders took the extraordinary step of jointly writing to the London *Times* just before Christmas 1940. Their letter on the "Foundations of Peace"[2] concluded with reflections on how, in peace, the world should arrange its political policy and social life. They proposed five standards "by which economic situations and proposals may be tested." The fifth test was as follows: "The resources of the earth should be used as God's gifts to the whole human race, and be used with due consideration for the needs of the present and future generations." Their letter made no dramatic impact, but it was a start toward proposing explicitly that we should manage the world not only for our own benefit but also for that of future generations. William Temple, as Archbishop of Canterbury, used the same language again at the Malvern Conference in 1941 and in a book in 1942. His language, probably coincidentally, turned up again exactly fifty years later when governments came to define sustainability as their goal in 1992.

Meanwhile, advocates began to extol, variously, the mystic underpinning of agriculture, the "back-to-nature" movement, and what is now called "organic agriculture." One of the more influential of these "new thinkers" was Aldo Leopold (1887–1948) who worked in the United States Forest Service and then as a university professor. As an advocate for a "land ethic" he wrote:

> Acts of creation are ordinarily reserved for gods and poets, but humbler folk may circumvent this restriction if they know how. To plant a pine, for example, one need be neither god nor poet; one need only own a good shovel. By virtue of this curious loophole in the rules, any clodhopper may say: Let there be a tree – and there will be one.
>
> If his back be strong and his shovel sharp, there may eventually be ten thousand. And in the seventh year he may lean upon his shovel, and look upon his trees, and find them good.
>
> God passed on his handiwork as early as the seventh day, but I notice He has since been rather noncommittal about its merits. I gather either that He spoke too soon, or that trees stand more looking upon than do fig leaves and firmaments.[3]

Leopold was a co-founder of the (US) Wilderness Society in 1935. To this day, The Wilderness Society expresses its mission in a way that quotes Leopold: "We are advocates for the land. At the heart of the work we do is the land ethic, which defines a set of principles in how humans should relate to the land"[4] (Fig. 55). Coincident with emerging concerns about the future of society and our management of land was the work of a disparate group of intellectuals who gave rise to "organic agriculture." Organic agriculture is, some would say, the most radical choice from among the "multiple futures" of agriculture and human landscapes. The organic movement is concerned with how we grow food, the extent to which we modify our land resources, how we process food, and consumer choices in buying produce. Advocates of organic agriculture might argue that it has been a core force in creating the paradigm of sustainability and allowing it to be embraced by affluent societies. However, without detracting from the development and dissemination of holistic or

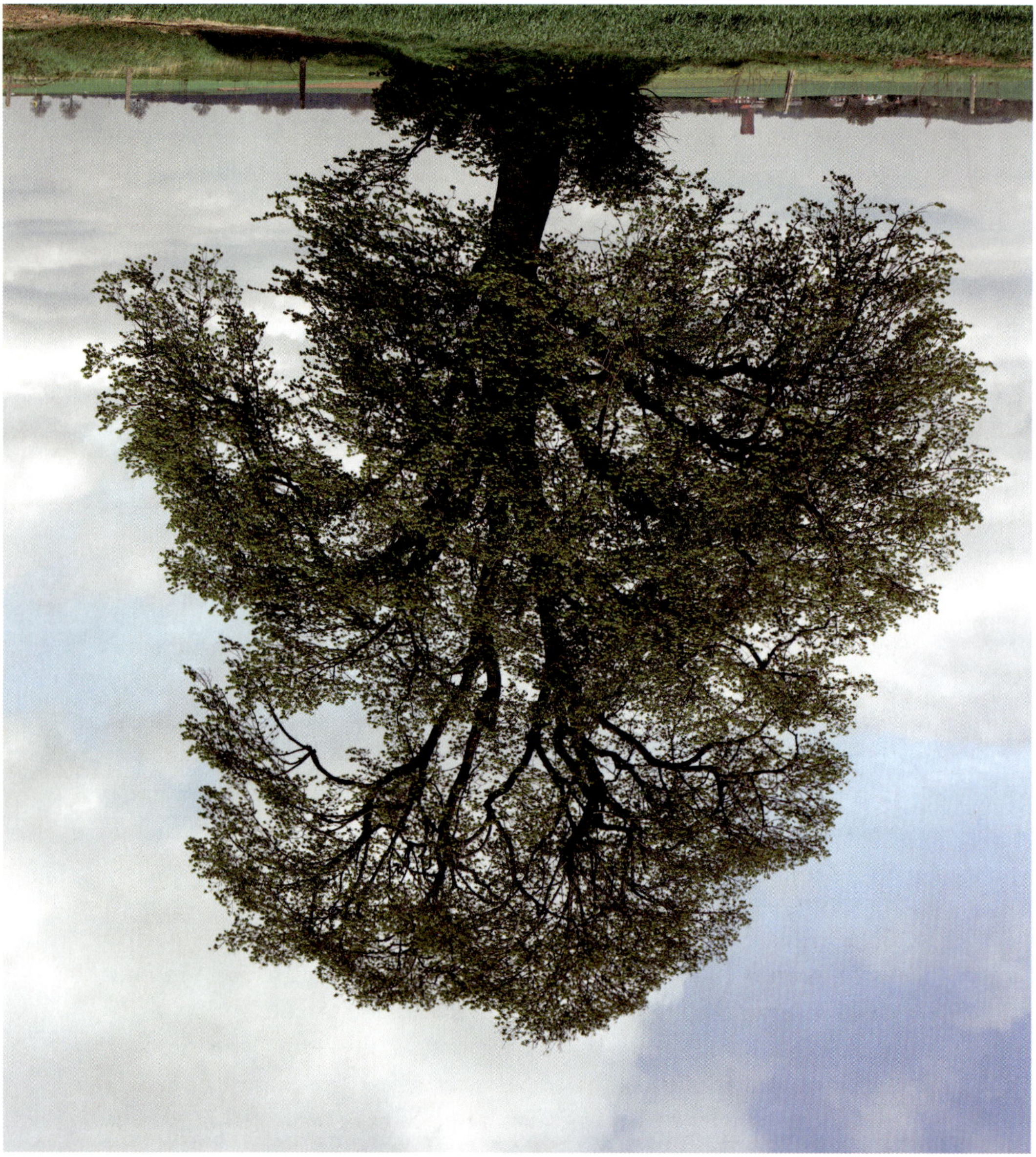

Figure 55
Rodney Graham, *Napoléon Linden, Marbais*, 1992–93 (C-print). Graham upsets our perception with his inverted tree, which seemingly grows downward from the sky. Viewing nature from a different perspective, we are led to reassess its value.

organic principles of farming, it is more realistic to say that organic agriculture has been a fellow-traveller: a small contributor to, rather than the creator of, the paradigm of sustainability.

The modern organic movement owes its genesis to a small number of innovator–philosophers who seem to have been isolated geographically and also to have been isolated from or ignored by mainstream agricultural science – at least until recently. Albert Howard (1873–1947), trained at Cambridge, spent his working life in India from 1905. It was only after he retired that he really turned to advocacy, and to travel, to promote ideas of composting and reliance on organic manures to enhance biological activity in the soil. His book on this subject, *An Agricultural Testament*, was published in 1940.[5]

Another key figure was Hans Müller (1891–1988), a Swiss politician who founded the Organic Farming Movement in 1930 and championed the autarky of the farmer and a more direct connection between on-farm production and minimally processed consumption. In the United Kingdom, Walter James, Lord Northbourne, was part of the "agricultural science establishment" as Provost and later chair of the Board of Governors of Wye College (now part of Imperial College), University of London. His book *Look to the Land*, published in 1940, used language half a century ahead of its time: "The soil is a whole world to itself. But as a world it is also an entity: a variable entity and a living entity, and one with which, as an entity, every farmer and gardener is intimately concerned. It is as a living thing, not as a dead medium, that the soil is most important to us."[6] Northbourne, who entitled a two-page section of his book "diversified organic farming," is credited as the first to use the term "organic agriculture" in its modern context; a slightly later book by Howard was subtitled *A Study of Organic Agriculture*.[7] Contemporaneously, Jeremy Rodale in the United States started *Organic Gardening* magazine in 1942, and Evelyn Balfour published *The Living Soil* in 1943.[8]

Why did all these pioneers publish in the midst of World War II when everyone's attention was elsewhere? Balfour was a member of the group that founded the Soil Association in the United Kingdom in 1946. The Soil Association lists the principal concerns of this group as: "The loss of soil through erosion and depletion, Decreased nutritional quality of intensively produced food, Exploitation of animals in intensive units, Impact of large intensive farming system on the countryside and wildlife."[9] For the first thirty years, the Association was based on a farm in Suffolk. There, they divided the farm's management into three units: one using the "new intensive" techniques, which today we would call organic agriculture; one farming traditionally (using inorganic fertilizers and sprays); and one using a mixed system. At the end of this period the results were not as clear as anticipated. This ambivalence in results – and the need, in most cases, for several years of transition before a system stabilizes in an organic mode – bedevils comparisons between organic and conventional agriculture even today.

Organic farming has been described as "both an agricultural philosophy and a farm management system. [It] is not going back 50 years in agricultural research [but uses] the very latest in technology, applied to current research, to fulfill the principles of good soil husbandry our forefathers adhered to."[10] The organic farm management system places importance on organisms, particularly those that live in and contribute to the fertility of the soil. Adherence to the principles of organic agriculture, as distinct from the broader (and variously defined) paradigm of sustainable agriculture, requires reliance on organic and slow-release fertilizers such as manures and crushed rock phosphate, rather than inorganic or artificially produced fertilizers such as John Lawes' superphosphate. Similarly, there are various "do's and don'ts," such as the avoidance of pesticides, which determine whether a farm or food can be certified and sold as "organic."

A key study among the hundreds that have since investigated various farming systems in the pursuit of sustainability was initiated in 1979 in the Netherlands. Although they do not use the term sustainability, the authors of a report on this experiment acknowledge that it was undertaken in response to public protest and opposition to the increasing use of chemicals in agriculture; this opposition had gathered momentum in Western Europe in the 1960s. The study compared, at great expense, three farming systems: a conventional system, with high levels of inputs, including pesticides; an integrated farm, with modest inputs, reduced yields and acceptable financial outcomes;

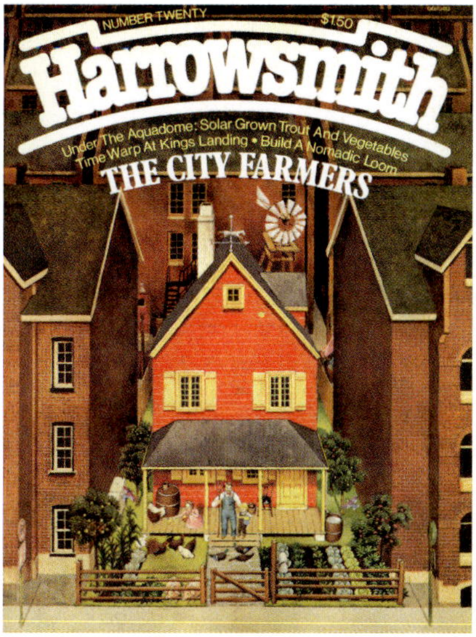

Figure 56
Harrowsmith, No. 20, July 1979. Growing interest among the public in organic agriculture has been reflected in magazines such as this one, championing the enthusiasm for the back-to-the-land movement with articles that promote an ecologically sound life, organic gardening, and city garden plots.

and an organic farm, using recycling and avoiding artificial fertilizers and pesticides.

The first five-year comparison of the three Dutch farms yielded mixed results. In his report on this long-term experiment, Zadoks[11] was unable to show clear-cut benefits in one system over another, just as Balfour had experienced thirty years earlier. Although each farm cultivated about the same area (some 17 to 22 hectares), the organic system caused soil organic matter and biota to increase most. The quantity of soil earthworms and mycorrhizae (root fungi) were 10 times and 2 to 3 times higher, respectively, in the organic farm soil than in either of the others. On the other hand, the organic farm had high labour costs and lost a great deal of money – although, cautiously, the scientists reported that "the small size of the farm does not allow any generalized statements to be made about the profitability of organic farming."

Although the debate about the profitability of organic agriculture, and the question of whether it is a system suited for niche or mainstream agriculture, continued into the 1990s, it is apparent today that, whether run under strict certified conditions or not, organic agriculture has a place and contributes to sustainability in contemporary agriculture. There was also a growing interest among the public reflected in magazines such *Harrowsmith* (Fig. 56), which helped to popularize the values of organic agricultures.[12]

Having taken an excursion into organic agriculture, we return to the church and the development of public opinion in the 1940s and 1950s. Exhorted by the church to think about future generations, some agriculturists, mostly on the margin, advocated organic farming, while other leaders pronounced pessimistic views about societies' attitudes and abilities to value natural resources. One such leader was Albert Schweitzer, a doctor, missionary, and concert-level organist who built and worked in a small village in Gabon, Africa, and was awarded the Nobel Peace Prize in 1952. Bearded and looking rather like God in Michelangelo's painting in the Sistine chapel, he would return periodically to Europe to make thunderingly pessimistic announcements: "Man has lost the capacity to foresee and forestall. He will end by destroying the earth."[13] This pessimism is seen in some contemporary art works (Fig. 57).

The philosophical and theological perspective – the need to value all life – espoused by Schweitzer converged with analyses carried out by groups such as the Club of Rome, a self-appointed, non-governmental organization whose membership is drawn from all regions of the world. Its "active members" are limited to one hundred scientists, economists, and business leaders. In 1972 a subgroup of the Club of Rome wrote a widely publicized report entitled *The Limits to Growth*,[14] which made projections on the speed with which the world is using up its natural resources.

To these voices were added the views of a minority of economists, who called for an economic valuing of natural resources to encourage their use as scarce, and not valueless, commodities. *The Limits to Growth* has since been criticized on a number of counts for inadequately anticipating new technologies and for setting its population forecast too high: it guessed that the world's population would reach 7 billion by 2000, whereas it merely almost doubled to 6.1 billion! However, the Club of Rome did achieve what the 1940 letter to *The Times* did not: it created a significant impact that attracted the attention of the world's politicians.

Fuelled by the Club of Rome's report, sustainability was the subject of academic writing in the 1970s and 1980s and rose to public prominence at a meeting in Rio de Janeiro in 1992. The sustainability of a system may be defined as the degree of difficulty we en-

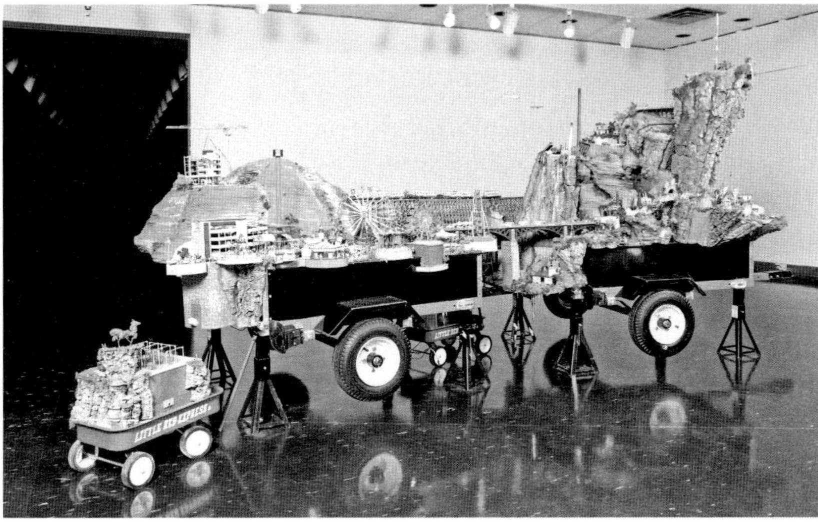

Figure 57
Kim Adams, *Earth Wagons*, 1989–91 (HO model parts, utility trailers, plastic landscape material, electrical components). Although earlier artists painted pastoral rural scenes featuring contented shepherds and cow herders, contemporary artist Kim Adams presents a completely different view. His remarkable sculpture attempts to compress a world view onto the top of a small utility trailer. The sculpture is liberally peopled with hundreds of miniature figures, houses, bridges, tunnels, cars, trains, airplanes, and industrial plants. The sculpture suggests a situation in which human ingenuity has gone wild as the density of population and industry threatens to collapse upon itself. The idea of this miniature world being transported on a wheeled trailer suggests that remnants of our world are being saved. The sculpture is endlessly fascinating, as it provides little vignettes, narratives and parodies of living and work spaces, and ingenuous constructions, such as the railway boxcars stacked to look like high-rise apartment buildings. Human habitation takes over the mountain top because all available farmland has long ago been devoured for industry and housing.

counter, or the amount of environmental, social, and economic resources we expend from outside, to maintain a community or ecosystem.[15] Hence the paradoxes above: as agriculture has moved from small-scale manorial or peasant-based systems to large-scale, industrial systems, its sustainability, thus defined, has generally declined; yet those in influential positions during the height of the paradigm of productivity would argue that we could not feed the world's population had we not moved to larger, "leakier" systems that are more dependent on external inputs.

"Sustainable development" was defined in the context of public policy in 1987 as development that meets the needs of the present without compromising the ability of future generations to meet their own needs. This, incidentally, echoes the earlier pronouncement from William Temple and his colleagues in the church. Policy for sustainability was further strengthened by governmental adoption through the "Rio Declaration" in 1992 of the "precautionary principle" with respect to possible environmental damage, loss of biodiversity and the like. The precautionary principle states that "lack of full scientific certainty shall not be used as a reason for postponing cost-effective measures to prevent environmental degradation."[16] It is an appropriate paradigm for the times, one that arose while the world's population increased from 2.5 billion in 1950 to 6.1 billion in 2000 and the state of natural resources generally deteriorated.

Achieving sustainability involves assessing a system or activity in terms of its biophysical (or environmental), social, and financial effects and requirements. It is a tall and complex order, which helps to explain why there have been few quantitative, multi-dimensional studies of sustainability and why, when society is debating whether to build a hog farm near a country town, or to let a city spread across a rich agricultural region, there are usually no numbers that take account of all these factors and on which we can base our judgments. The absence of data, and the general difficulty posed by political debate involving conflicting views and beliefs, creates a situation in which, despite a general endorsement of the concept of sustainability, there is no agreed method by which to define sustainability in a given situation. There is no single, unified view of the agricultural landscape (Fig. 58).

It may thus come to pass that sustainability, unlike productivity, will be more useful as an aspiration that creates a platform for debate, rather than as a measurable goal that industry and governments can pursue with the single-mindedness that characterized the era of productivity. Nonetheless, it is often easy to see when things are not sustainable – as in the hog farms described in the previous

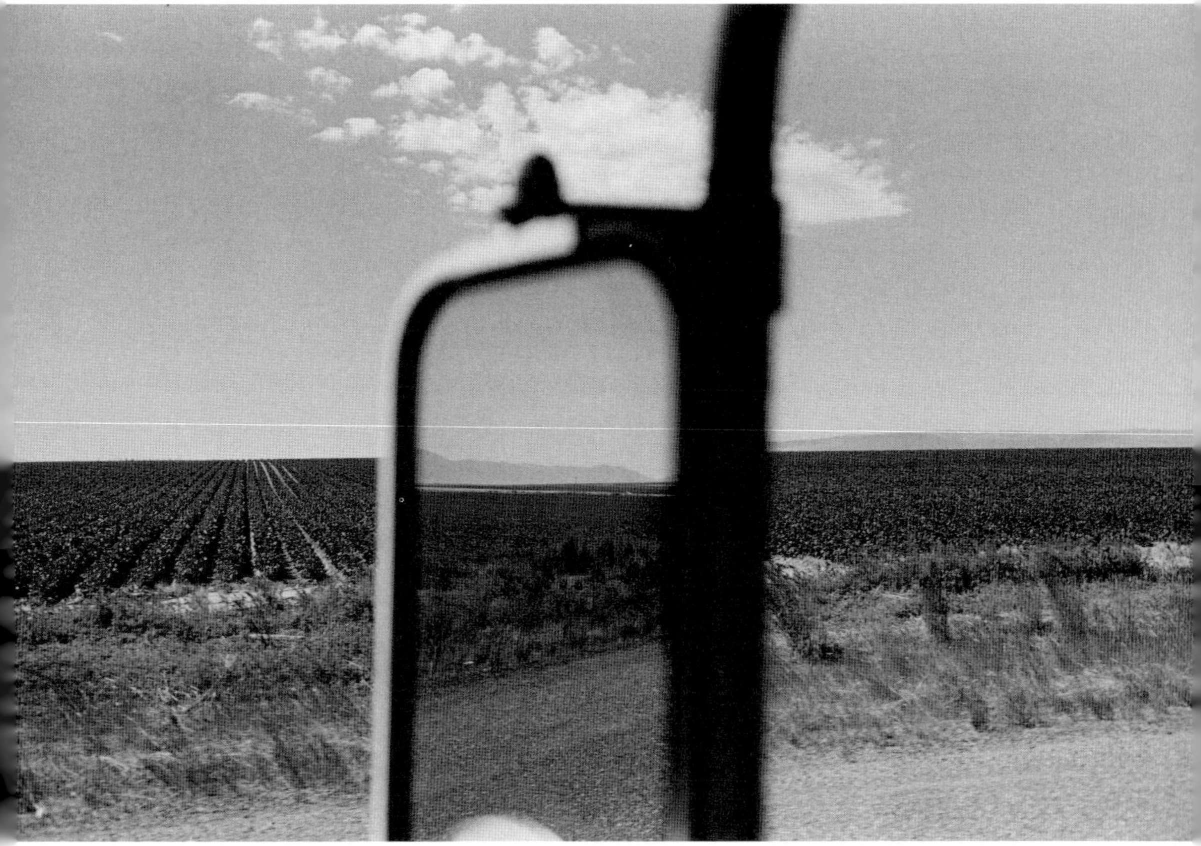

Figure 58
Lee Friedlander, *Idaho*, 1972 (silver gelatin print). Lee Friedlander transformed American documentary photography in the 1960s, capturing the "social landscape" in multiple series of photographs. Friedlander's *Idaho* depicts the uninterrupted farmland, an utopian view broken only by the side-view mirror that reflects back the flat horizon behind the camera's lens. It is through Friedlander's aesthetic choices that this otherwise idealist image is shown as fragmented.

chapter, or villages in the sub-Sahara, or the highlands of Ethiopia. These agricultural systems characteristically do not invest, or reinvest, in natural resource maintenance.

What does a failure to reinvest in natural resource maintenance look like? In Ethiopia, cropping has been a way of life since before recorded history. It is one of the centres of origin of modern crops, which is taken as evidence of a long, uninterrupted civilization based on agriculture. The season in the Ethiopian uplands begins with plowing, using a wooden, single-furrow plough that would have changed in the last two thousand years only insofar as some of the leading edges are now made of steel, not wood. The crops, sown by hand, are also harvested by hand, and the grain is threshed from the seed-heads and straw with simple machines. In more remote areas, oxen quietly walk in a circle, knocking the grain onto the compacted, swept ground with their hooves. Once the harvest is made, some of the straw is bundled and taken for storage to feed cattle. Scientific agriculture has documented, many times over, that the straw is so low in nutrients that it will not maintain the body weight of cattle, or give enough nutritional support for them to maintain milk production at what a Western farmer would consider a reasonable level. Nonetheless, it is the best and the only mobile source of energy, nitrogen, phosphorus, and other nutrients available. Sheep, goats, and a few skinny cattle also graze across the fields, picking up the grain dropped before harvest, or the straw left attached to the roots and soil.

Toward the end of the dry season, as the humidity rises and the time for rain and planting draws near, the farmers pull together piles of remaining debris, scratching roots and stalks from the field into cairns about fifty to eighty centimetres high and two metres across. These are then burned; fires are painstakingly started and restarted as the cairns glow across the fields in the night. Soon there is little left of the organic matter, but the mineral nutrients, such as phosphorus, are back in elemental form, mixed with the soil. The ash is then raked across the field, and the plowing and planting cycle begins again. In this way, about two-thirds of the nutrients in the crop are removed with the grain each season, and the remainder that stays in the field is made quickly available for the next crop. Although the burning and spreading-out of nutrients is a good idea, the annual removal of much through grain, and lack of re-investment through the use of compensating fertilizers or nitrogen-fixing manure crops, means that the soil becomes less and less fertile.

Turning the concept of not reinvesting in natural resource maintenance to a more positive goal gave rise to "ecosystem health." This is a useful term, to help us think about what is acceptable and what is not acceptable in this regard. Three attributes of ecosystem health have been identified: vigour (productivity); organization (generally, the complexity of the organism chains within the ecosystem); and resilience. As Rapport and colleagues write:

> Vigor or productivity refers to the capacity of the system to sustain the reproduction of both plants and animals. Organization refers to the capacity of the system to support a diversity of lifeforms and their interactions. Resilience refers to the capacity of the system to buffer perturbations; that is, the capacity to rebound after disturbances such as fire, floods, windstorms and the like. Although these attributes have been worked out in terms of the biophysical properties of the systems (that is, from ecological perspectives), the concepts also apply to the socioeconomic and human health dimensions of ecosystem health. For example, in a healthy ecosystem, economic activity is buffered against the vagaries of market forces, for the system can support a variety of alternative human activities that can be brought into play to maintain a source of incomes for the human communities within the system.[17]

Unhealthy or degraded systems, from an ecological viewpoint, show characteristic symptoms. These are most apparent as a reduction in biodiversity. This is thought to be an early warning sign, as in the loss of microbial species and other small organisms in the soil as it is over-cultivated and compacted. Other signs are declining productivity: that is, declining crop yields resulting from the degradation of soil structure or from salinity; disruption and leakage of water or nutrients; and the shift from long-lived biota to short-lived species. All these can be amply illustrated by the onset of salinity: soil animal populations plummet, and long-lived trees grow more slowly and then die, to be replaced by annual grasses and exotic (foreign) short-lived species, which in turn die. All of this occurs while the system becomes more and more "leaky": more and more of the rainfall and soil nutrients are washed out into the waterways, thereby spreading the problem.

In our complex world, agriculture may often contribute to ecosystem stress and declining sustainability. However, it is seldom the only contributor, and often is not even the main one. Consider two areas: the Great Lakes system in northeast North America and the Kyronjoki estuary of southwest Finland, which flows to the Baltic Sea. Rapport and Whitford[18] contrast these systems: the Great Lakes Basin has been restructured by drainage, dredging of harbours, removal of shoreline vegetation, canals, and regulation of river flow through dams, bank-works and barrages; the Kyronjoki has been exposed to much the same engineering "improvements." Both suffer from nutrient-rich runoff from agriculture and sewage.

In the first case, it might be argued that extensive clearing over two hundred years, particularly for logging and agriculture around Lake Erie and Lake Ontario, have impoverished habitats and affected the hydrology of the region. A reduction in area of a forest is likely to cause a loss of species that live in its interior. A smaller area also increases the "edge effect," the length of the boundary and the distance from the core, which in turn affects the likelihood of invasion by exotic plants or animals. Agriculture has exacerbated this degradation by keeping the flora simple and causing nutrient runoff, which adds to the pollution from industries and a highly urbanized population of some thirty million people within the catchment. In the case of the Baltic estuary, the same applies: highly seasonal river flows encouraged catchment control and drainage that, because the soils were acid sulfate, released sulfuric acid when they were waterlogged. The resultant toxicity denuded vegetation that then increased the severity of runoff and nutrient leakage. Thus, the system worsens. Agriculture, along with urban waste, contributes to pollution while, as the primary rural land use, it both inherits and appears to be the leading cause of landscape degradation.

The interconnectedness and fragility of sustainability – the knife-edge that separates a functioning catchment from a nonfunctional one – was illustrated for the Great Lakes region in August 2003, albeit by a malfunction in infrastructure rather than of the underlying biological system. A power surge in the electricity grid in Ohio was transmitted throughout the system. Unchecked, it caused fail-safe systems to activate in hydroelectric and nuclear power plants throughout the northeast region, shutting down much production. For as long as fifty hours, fifty-five million people had no electricity.

What has happened in the Kyronjoki estuary is also happening across large areas of Brazil, where acid sulfate soils have been cleared for food production since the 1970s. What has happened in the Great Lakes region is also a case study for Europe as a whole, where deforestation and high levels of fertilizers leaking from agriculture, along with industrial wastes, are causing groundwater, rivers, and the air to become contaminated. This effect also closes the circle on the deforestation that continues to take place in South America (described briefly in chapter 2). Here, it is important to consider not only the loss of forest area, but also the fragmentation that occurs through partial clearing. This fragmentation creates small pieces of forest that are susceptible to loss of species and large edge effects; between 1978 and 1988 alone, these fragments increased from 208 to 588 thousand square kilometres as the total area of forest fell.[19]

The European experience in the last decade indicates that as societies adopt the paradigm of sustainability, they embark on a learning process. Although one-dimensional sustainability is not enough, it is better than nothing, and it may lead to the adoption of land-use practices and to the design of landscapes that do not compromise the ability of future generations in a genuinely multi-dimensional way (Fig. 59).

A core issue affecting global agricultural sustainability is the level of price support for farm production in Europe and the United States. These supports are possible because they are provided by very affluent, urbanized societies, whose wealth is created largely in the industrial and tertiary or service sectors, rather than in the agricultural sector. The supports are enormous, and actually exceed the dollars that several rich countries devote, with some fanfare, to aid for developing countries: the European Union subsidized agriculture to the tune of $5.98 billion per year (1999 data), and Switzerland, Norway, and the United States all heavily subsidize and protect their domestic agricultural production. The twin supports of subsidies for production and tariffs against food imports effectively promote in-

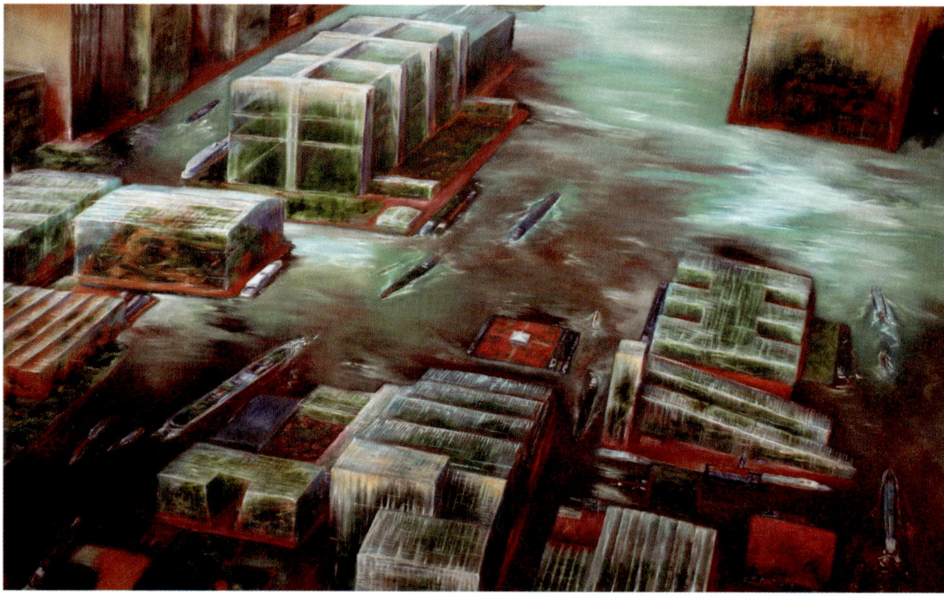

Figure 59
Eleanor Bond, *Activity in the Inner Harbour is Regenerated by the World Botanical Garden, Constructed with Recycled Materials from the Glass City*, 1995 (oil on canvas, unstretched). Artist Eleanor Bond asks the question: Will our farmland be so degraded in the future that we will be forced to feed ourselves with plants grown in artificial environments? She creates a dystopian view of how junk from the megacity will be used to create endless greenhouses in derelict industrial sites. Her painting also acknowledges our human ingenuity in finding new technologies to solve our problems.

tensive agriculture and high levels of production. A positive outcome is a highly managed rural landscape and lively country towns. A negative effect is that it does nothing to enhance the competitiveness of low-income countries, and thus agricultural production on a global scale becomes distorted and, in some regions, slips into nonsustainability.

Another dark corner of this policy is that it causes, or at least permits, local environmental problems in high-income countries. High prices for wheat and other commodities reward the high usage of fertilizers such as nitrogen on cereal crops, and condones high levels of waste such as manure and methane from high densities of intensively managed livestock. Thus, current financial support for agriculture in some high-income countries creates a two-fold damnation, affecting both their own environments and the well-being of rural areas in low-income countries.

Attacks on price supports within the European Community (EC) and the United States have been waged more or less continuously by countries outside these blocs through the General Agreement on Tariffs and Trade and its successor, the World Trade Organization. However, progress toward sustainability is most likely if it begins at home. As early as 1991, the EC Agricultural Commissioner, Ray MacSharry, proposed reforms of price reductions, supply controls, and direct compensation to farmers, which would have reduced agricultural production, perhaps by 50 per cent, in the EC.[20] This was not accepted. Subsidies and production continued to be maintained, and in some cases rose, throughout the 1990s in Europe, the United States and Japan.

The billions of dollars shifted from urban Europe, the United States, and Japan into their agricultural sectors are mind-boggling. They can be categorized into two boxes: market price support and direct payments. The market price supports in the late 1980s amounted to $15, $15, and $32 billion per year in the European Union, United States, and Japan respectively; these subsidies subsequently dropped for the two big trading blocs but rose in Japan, reaching $10, $9, and $43 billion respectively in the year 2000.[21]

As for the second category, direct payments to farmers, these have risen in all cases and are on the same scale, or greater, as market supports. Among the big-ticket items of subsidization, milk production in the European Union received about $19 billion each year in the 1980s and 1990s, falling to a mere $13 billion recently; in the United States, the market support for dairying was $12 billion, falling to $9 billion; and, in Japan, price support for domestic dairy produce has held at about ¥600 billion per year throughout the era.[22]

Instead of addressing the subsidy-fuelled causes of excessive domestic food production, the EC embarked on a journey to address a symptom of nonsustainability, namely pollution. This will be described shortly. The structural problem of artificially promoted intensive agricultural production would not go away, however. Some reform became inevitable with the expansion of the EC to twenty-five countries and was foreshadowed by positive referenda in the Czech Republic, Poland (which has as many farmers as France and Germany combined), and other parts of Eastern Europe in mid-2003. The "reform" of the price supports embedded within the Europe-wide agricultural policy was announced in June 2003.

At the same time as the EC was distorting trade and prices with high levels of domestic agricultural production, it sought to address the issue of sustainability by taking steps to reduce agricultural pollution. In 1991 the Council of the EC adopted the "Nitrates Directive," more formally called Council Directive 91/676/EEC,[23] "concerning the protection of waters against pollution caused by nitrates from agricultural sources." This aimed (or aims, as it is still being implemented) to reduce pollution from agriculture through excessive use of inorganic fertilizers and the spreading of animal manures. The Nitrates Directive followed directives dating to as early as 1975 that were concerned with the quality of human drinking water. The Directive required member states to identify water that was polluted or at risk, when nitrate levels exceeded fifty milligrams per litre. Once this was done, the agricultural areas that emptied into these waters were to be designated as "vulnerable zones." This was to take place by the end of 1993 and was to be updated every four years.

Having designated vulnerable agricultural zones, mandatory measures were to be implemented to reduce the impact of agriculture on water quality. The main measures were to ensure: (1) that every farm should have adequate storage for livestock manure so that it could regulate the spreading of the nutrients and not be forced to unload manure at times when it would not be absorbed in the soil; (2) that manure must be limited to no more than 170 kilograms of nitrogen per hectare per year; and (3) that the application of fertilizers should be based on a calculated "nutrient balance."

The notion of a nutrient balance is basic to scientific agriculture. It calculates the amount of fertilizer that may be applied from all sources (livestock manure and inorganic fertilizer from bags) and not exceeded as that which is taken up by the plants and subsequently harvested. This can be calculated according to how big the crop grows and the nutrient concentrations in the dry matter of the various plant parts. A critical refinement is to also recognize that the uptake by the crop depends on its size and rate of growth, so that, in addition to calculating the total amount of fertilizer that may be applied during the life of a crop, it is also necessary to estimate how much fertilizer can be taken up at any one time and, if necessary, to split the applications across the life of the crop. Clearly, while the idea of nutrient balance is simple and fundamental, it was obviously exceeded in practice by over-enthusiastic farmers and grasping fertilizer salespersons; otherwise, the problems we see in Europe and parts of North America would not have arisen.

The sentiments that caused the Nitrates Directive to be adopted were good; however, the practice falls short. Only Denmark and the Netherlands identified vulnerable agricultural zones within a few months of the deadline, and a decade after the directive was made, some countries (such as France and the United Kingdom) had yet to implement their action programs. There are a number of reasons for non-compliance, on both sides: the agricultural industry has lobbied against reforms whose costs would put some producers out of business; the directive itself is in part defective; and technology and implementation processes have, to date, been inappropriate. It is "simply impossible to run a top-down control system for some 100,000 farmers (in the Netherlands) that is based on highly detailed paper forms that have to be sent to a central office to be processed."[24]

Despite the shadow that has passed between the aspirations of the directive and its implementation, it has nonetheless brought attention to environmental sustainability and the key role of agriculture in achieving this. It has caused monitoring to increase, and maps to be drawn. It is asserted, for example, that some countries – those with high livestock densities, notably Netherlands, Belgium, and Denmark – exceed the "balanced nitrogen budget" by hundreds of kilograms of nitrogen per hectare per year.[25] These excesses on a national scale must conceal massive over-fertilization in some local areas.

It has also been calculated that although agriculture contributes 45 per cent of the nitrogen and 20 per cent of the phosphorus to three of the main waterways – the Rhine, the Elbe, and the Po – the total pollution from agriculture has indeed declined slightly.[26] Two other findings are noteworthy. First, although the reduction in nutrient waste into the Rhine and Po was small, there was a massive decrease in nutrient use in agriculture in Eastern Europe. Regrettably, this cannot be attributed to an awakening among farmers or governments to the ideal of sustainability. Rather, it followed the collapse of communism. Former Soviet farmlands now lie fallow, large-scale processing factories are crumbling, and villages are occupied only by pensioners. This breakdown of agriculture will have a significant beneficial effect on the health of the Elbe.

Second, there is good reason to believe that the directive will not have sufficient impact. Some calculate that, if it is implemented, it will cause agriculture to reduce the total nitrogen load in European rivers by only 20 to 30 per cent. It is also likely that, if farmers act rationally, the blanket annual target of no more than 170 kilograms of nitrogen per hectare will encourage livestock manure spreading to shift to areas where historically it has been less – and perhaps for good ecological reasons, such as the unsuitability of the soil. Even now, animal manure is being taken from the Netherlands to be spread in Germany. These problems are not insurmountable, but they do suggest that a major change in landscape management in Europe will require more sophisticated, regionally based, national legislation. It will need to take into account soil types and soil capacity to absorb nutrients, and the fact that industrial change, such as the relocation of intensive or industrial livestock farming, will depend on political will and market prices.

The first international adoption by policy-makers of the paradigm of sustainability was at the "Earth Summit" in Rio de Janeiro in 1992. Then, European nations collectively tried to implement this paradigm by addressing one particular symptom of nonsustainability, namely nutrient pollution of waterways. It could be argued that, in introducing the Nitrates Directive in 1991, social concern joined environmental concern and monitoring. However, it is simpler to see the subsequent actions as environmentally driven. The Directive has not yet had its intended impact, probably because it has not simultaneously addressed the social and economic aspects of the same problem. Where are the social processes – perhaps, the participatory

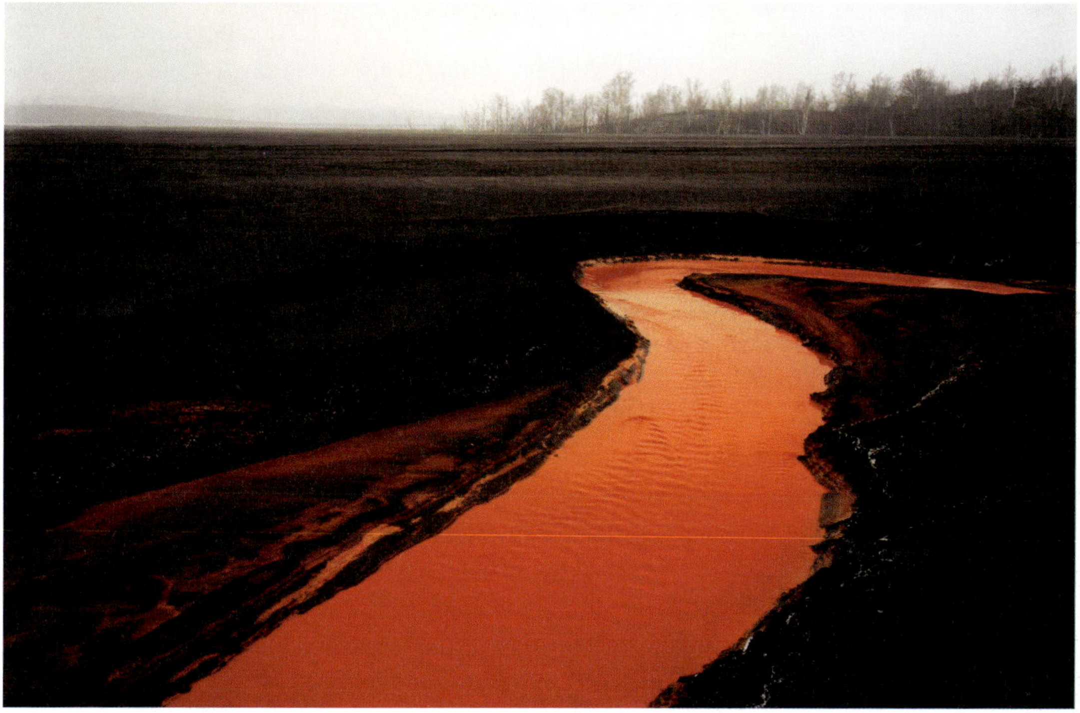

Figure 60
Edward Burtynsky, *Nickel Tailings #34, Sudbury, Ontario*, 1997 (chromogenic colour print). Burtynsky's photograph, with its lurid orange colouration, documents a stream carrying tailings from nickel mines in Sudbury, Ontario.

farmer-led action – that would create ownership and commitment and bring the Directive into the mainstream of peoples' livelihoods? Where are the economic incentives – carrots and sticks – that would cause industrial agriculture to scale back or relocate? The existing environmentally driven legislation in fact has perverse consequences: livestock will remain in place, and manure will be transported to other countries, or else that manure will remain, supporting an increase in cropping. After all, the allowed spreading of 170 kilograms of manure per hectare relates to hectares of cropland. This discussion, of course, does not address a water-quality issue that works in the opposite direction: industrial pollution, which can poison water sources for livestock and make soil unsuitable for farming (Fig. 60).

Sustainability, though, is broader than muck and manure, organic agriculture, or water quality. It is about the "triple bottom line" of trading off needs in the biophysical, social, and economic spheres. The logical end point of purely environmental considerations is to retain or convert the world into a national park that supports environmental objectives, such as biodiversity. But it does not feed the current world's population, nor does it maintain people in rural landscapes. Principle 1 of the 1992 Rio Declaration on Environment and Development states: "Human beings are at the centre of concerns for sustainable development. They are entitled to a healthy and productive life in harmony with nature."[27]

Although the European Commission was attempting to mandate environmental sustainability with respect to water quality, others have given attention to social and economic sustainability. The early United Nations Human Development Index was a compilation of three social and economic indicators at a macroeconomic level. It considered life expectancy at birth, educational attainment, and standard of living in terms of gross domestic product per capita. These were mathematically standardized to calculate a single number, a Human Development Index for each country, a pecking order of human prosperity.

More recently, the Davos World Economic Forum (2001) produced an Environmental Sustainability Index[28] for 122 countries. This was based on twenty-two core "indicators" chosen to address five things:

- Environmental systems. A country is environmentally sustainable to the extent that its vital environmental systems are main tained at healthy levels, and to the extent to which those levels are improving rather than deteriorating.
- Environmental stresses. Human-made stresses such as land-clearing should be low enough not to cause demonstrable harm to the environment.
- Human vulnerability. Sustainability requires that basic human needs such as heath and nutrition are met.
- Social and institutional capacity. Effective responses to challenges, and progress toward greater sustainability, require

that countries have institutions, skills and attitudes that will work "in the right direction."

- Global stewardship. Points were awarded to countries that demonstrated global awareness through international treaties and measures such as greenhouse emissions and chlorofluorocarbon consumption.

The Environmental Sustainability Index database includes the amount of land that has been cleared for agriculture (as monitored by satellites) and the severity of human-induced soil degradation. It also shows the use of fertilizers and pesticides per hectare, as well as biodiversity measures such as the percentages of mammals and birds at risk of extinction. The output again, is a single number per country: Australia's score is 70.7, Albania's is 44.2. One is good, the other bad. So what?

Although the Davos scores can be dismissed as a pecking order devised and justified by scientists from high-income countries, it is the numbers within the analysis that will be most instructive and that, one hopes, will open a path to the future. For example, despite its "good" sustainability index, Australia scores worse than Albania in the agricultural rating included under the heading of "reducing environmental stresses." Also, when the Davos group analyzed the inputs most highly correlated with their country-aggregate sustainability index, they found three that carry a message for everyone. Those three key indicators help to explain why some countries are creating more viable agricultural sectors than others, and why the Green Revolution (chapter 3) was more effective in some countries than others. They are: lack of corruption; stringent government regulations relating to the environment; and scientific and technical activity as measured, crudely, by science articles published per million people.

How has the paradigm of sustainability altered agriculture, our view of our selves, and our landscapes? The impact is limited, but there are important things to note in answering these questions. Perhaps it is a matter of time: the paradigm has coexisted for only ten, or at most fifteen, years with the paradigm of productivity. Also, it is only the idealists who currently believe that it is society's dominant paradigm. Although sustainability might now be uppermost in government rhetoric, most policy-makers (as distinct from elected politicians) in Western democracies have been trained as neo-classical economists within the paradigm of productivity. Also, productivity measured in economic terms is still a value closely held by the majority of farmers and their farm organizations.

All that being said, the concept of sustainability very importantly gives people a language that can be used to debate "progress." Part of that discussion about progress hinges on who makes decisions: Should government, or the community, have stewardship of agriculture and the environment? This is not a relabelling of the old farmers' cry for more taxpayer support because some of them are poor, face uncertain weather, or need subsidies to trade internationally. Rather, it recognizes that an agricultural practice – for example, planting a crop of corn, or building a shed to house broiler chickens – has effects outside the field or system that the farmer directly controls. The corn might have no such effects, or it might create externalities such as runoff of nutrients or increasing salinity that will affect the farmer's neighbour long after the corn is harvested. Similarly, the broiler shed will persist only if society maintains good roads for the transport of inputs (chickens, feed, etc.) to the farm, and of chickens and manure from the shed.

There are two schools of thought about stewardship. One simply says that, as actions have an impact beyond their immediate patch – the field on the catchment, the catchment on the region, the region on the country, the country on the world – then it is up to the "main actor" at the higher level to ensure careful decision-making at the lower level. Ultimately, at high levels, governments are these actors. Leave it to the government!

This is the rationale that has worked well for the management of cities in developed countries: municipalities or state governments set and support the rules that constrain individual rights and by which citizens abide to ensure, for example, that their garbage does not accumulate. Clearly, this system works less well in the burgeoning urbanization of less developed countries. The other thought about stewardship is that, particularly in non-homogeneous rural landscapes, it requires participatory decision-making, regardless of whether it will ultimately require government rules or not.

The participatory view is based on five considerations. First, city dwellers have accepted that they must fit within certain norms: for example, the placement of household garbage in a bin that is put on the street on only one day of the week, or the acceptance of certain losses of rights, such as the right to building domestic structures without a permit or to spray pesticides for "cosmetic" landscaping.

Farmers are more isolated and, as some psychological surveys show, have different values and ways of thinking than "average" urbanites. They are thus more concerned with personal freedoms, as witnessed by the gun lobby in the United States and by farmer protests in various countries against proposed legislation that might curb their freedom to clear land. On the other hand, if farmers meet and collectively decide that something is worth doing, then "participatory decision-making" overcomes the perception that the decision is imposed and helps to ensure compliance.

Second, stewardship of the landscape is based on a mix of personal values (Fig. 61) as well as on objective knowledge of such things as prices and returns and and environmental effects of certain farming practices. These values are not unrelated to history. For example, the immigrants who created the corn belt in the United States (and, further north, the Canadian prairies) were often from central and northern Europe. The landscape, the farms, and the roads were laid out on square grids, a section being a square mile (640 acres, or 259 hectares) and a quarter-section being each side of the square divided into two, without much reference to hills or streams. Today, these organized landscapes remain. The farmers take pride in how their crops look and the absence of messy corners or "wasteland." Neatly painted barns and houses are important, too. Unkempt strips, or the practice of leaving fallow those areas that do not yield well, may conflict with the values expressed in this cultivated landscape, even if nowadays these farmers know that such actions might assist the conservation of biodiversity. From the 1960s to 2000, corn-belt agriculture doubled the area of soybean cultivation and increased corn cropping too, at the expense of species diversity and activities

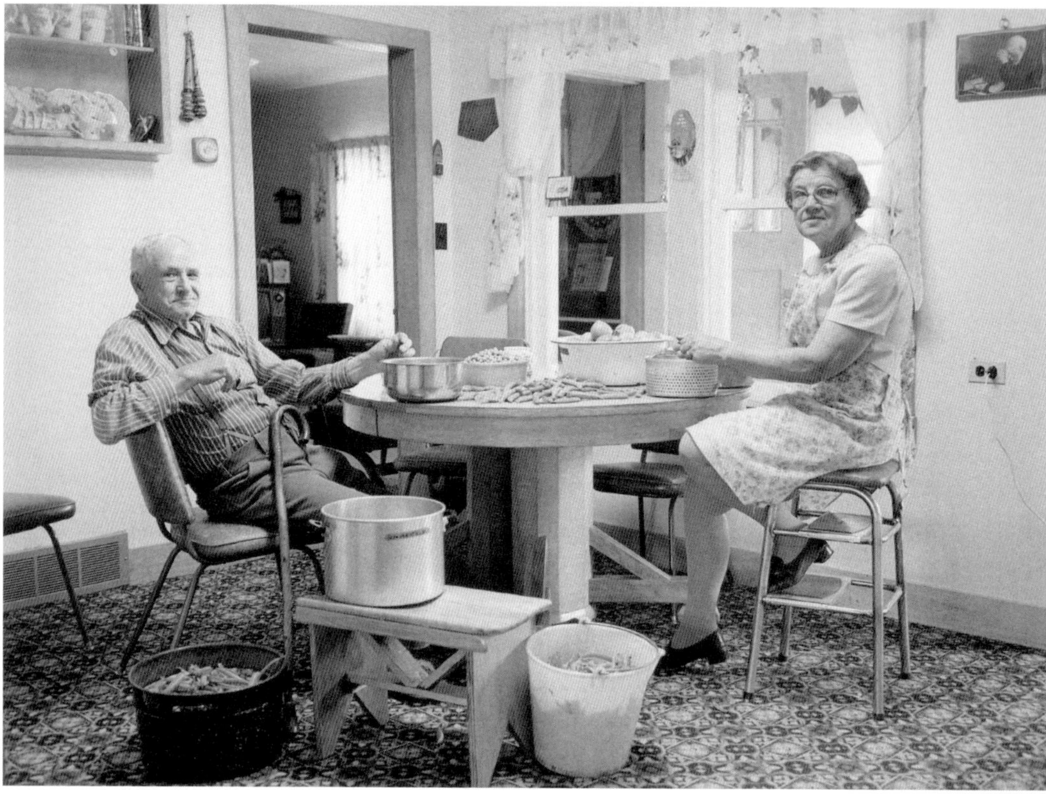

Figure 61
Orest Semchishen, *Glen and Hilda Cole, Near Coronation, Alberta*, 1980 (silver gelatin print). Mr. and Mrs. Cole contentedly process the fruits of their labour from their kitchen garden, continuing an annual family farm tradition.

such as dairying and pig farming. Although this trend may have been driven by the quest for productivity and specialization, it increased the neatness of the landscape. Changes in personal values, not government rules, will be necessary to bring back the long-lost short-grass prairie, its birds and other wildlife.

Speaking of birds reminds us of quite a different illustration of the importance of personal values in achieving sustainability. A development project sought to introduce tree legumes in the highlands of Ethiopia, where the soil was degraded and the perennial trees added to the nitrogen in the farming areas through the spreading of seeds and by the practice of cutting and feeding the leaves (rich in nitrogen) to livestock, which produced manure that was spread on the crops. An old farmer, sitting outside his mud-walled home, was asked about the major benefits of the program, and why he supported it so vigorously. He spoke not of shade trees or nitrogen but replied: "It has brought the birds back."

The third and fourth reasons for keeping policy-making and compliance at the level of the affected community are that governments are increasingly distrusted and do not "get it right." Lack of trust is exemplified by the British mishandling of the BSE and foot-and-mouth outbreaks in 1999–2000. Wholesale slaughter and restrictions of transport, after years of inaction, not only discredited the relevant policy-makers, but also made the public skeptical about government handling of disease outbreaks, a skepticism that was evoked again by the isolated case of BSE in Alberta and during the SARS (severe acute respiratory syndrome) outbreak in Ontario, both in 2003.

Local and regional differences, and the inherent complexity of sustainability, make it highly likely that a one-size-fits-all approach will not "get it right." The case of water pollution in Europe illustrates this. The development of one-size-fits-all indicators of sustainability, applied to different regions or systems, misses the point. Even the Kyoto negotiations for the mitigation of global warming recognize this by setting different country targets for the abatement of greenhouse gas emissions and, within each country, different priorities and government incentives to attempt to achieve these targets.

Governments' inability to "get it right" is made increasingly likely by their withdrawal of expertise from the issues. As food production, agricultural policy, and landscape design become increasingly important, Western governments have scaled back their employment and funding in relevant agencies. It is asserted that, in Canada, which has probably maintained government funding better than some, there has been a 37 per cent decline in person-years employed in public agricultural and food research, and a 55 per cent decline in government expenditure on technology transfer for the sector.[29]

Finally, skepticism about the effectiveness of government aside, there is the complexity of sustainability. Can any individual or group develop a policy that brings all the issues together? There is a momentum toward governments coordinating actions by others, whether by more provincial governments or by self-appointed gatherings of stakeholders in community action. To illustrate this complexity, consider the policy documents written by the Australian government since 1990; as if the difficult process of drafting these policies were not enough, they still required sign-off by state governments. Papers relevant to agricultural sustainability include: the National Forests Policy, the National Greenhouse Strategy, the National Strategy for Ecologically Sustainable Development, the National Strategy for the Conservation of Biological Diversity, the Council of Australian Governments' (COAG) Water Reform Framework, and the National Action Plan for Salinity and Water Quality.

Although community action is no doubt as old as communities themselves, government-supported community action is a more recent phenomenon that has paralleled, and supported, the paradigm of sustainability. It is best exemplified by the movement known as Landcare in Australia. Rural communities with common concerns had begun to organize themselves around the issue of land-clearing and rising salinity (Fig. 62). By the mid 1980s, groups of farmers – and some public servants working in regional conservation offices and agriculture agencies – had lost patience with the top-down bureaucratic approach to addressing these issues. And then an interesting development occurred in Warrenbayne-Boho, in the southeastern corner of the country.

The Warrenbayne-Boho hall could be the community hall in any one of countless Australian communities. In fact, it is not in a town, but sits on a rise at the side of the road, the local communities being too sparse to support a town or even a gas station. Wooden-sided and seriously in need of paint, it has a steeply sloping galvanized iron roof and a gravelly paddock for parking to one side. Inside, it is dark and hot: there never seems to be a breeze strong enough to clear through the hall, given the small windows and its situation on a hill, where eucalyptus trees cut the air movement, too. Here, more than a hundred farmers and a few outside supporters met and sat on stackable blue plastic-and-steel chairs to create what is generically called "a community of common concern" – regarding farm planning, tree planting, and salinity amelioration. They had no recogni-

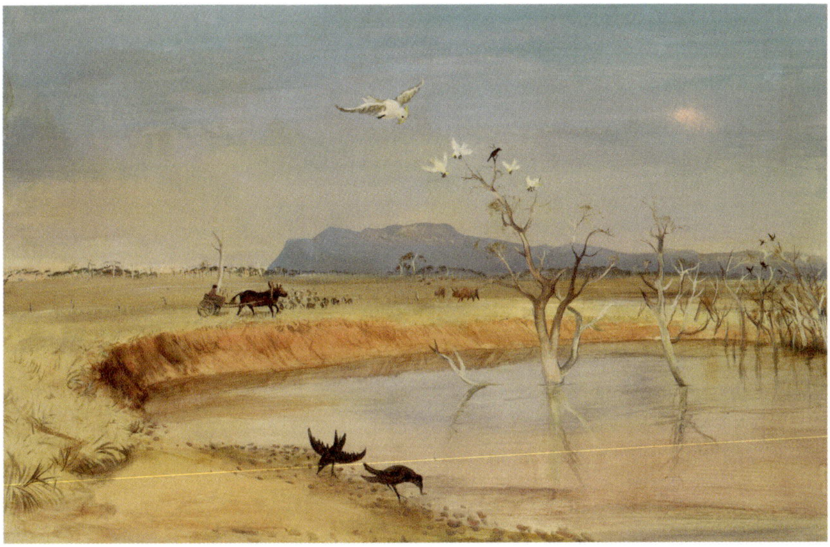

Figure 62
Arthur Boyd, *Irrigation Lake, Wimmera*, 1950 (resin and tempera on composition board). Arthur Boyd's painting depicts the fragility of the open plains and water containment in the Australian outback.

tion, but they were not alone: similar groups were meeting across the country to discuss similar issues.

Those who believe that a paradigm shift requires an alignment of the heavens, or a coalescing of people with "the right stuff," will find support for their theory in the birth of Landcare. No one knows what might have arisen, or not, solely from the community talk-fests occurring across the continent. But the fortunate alignment that did occur was the appointment in 1985 of Joan Kirner, an urban politician, as Minister of Conservation, Forests and Lands in the government of the State of Victoria. Kirner recounts that on her first day as minister she asked what was working well, and what was not. She was puzzled with the contradictory reply that the State Soil Conservation Service was going well, but land degradation was a big problem. Kirner's puzzlement created action; her philosophy provided the fresh direction that would create, and empower, community groups. Later, as Premier of the state, she summarized her philosophy: "Community development requires that the affected community participates in the decision-making, identifies the challenges, develops the solutions and owns the outcome."[30]

Kirner met with the few community groups that were in existence, including the Warrenbayne-Boho Land Protection Group. Within twelve months a government-sponsored community-based program to address what would be called sustainability was sketched out. The first Landcare group was launched at Terry Simpson's property near St. Arnaud, Victoria, in 1986. Here, things might have died a quiet death, but Kirner (who became Australia's first female state premier) enlisted as a partner in this effort Heather Mitchell, who become the first female president of the Victorian Farmers Federation. Together they mainstreamed the concept, creating the Victorian Landcare Program in 1989.

Then a second heavenly alignment occurred. Landcare was launched in Victoria while other states were dithering with more top-down catchment management committees and boards. But, at the national level, a new government was looking for ways or programs to lock in support from "green voters" or conservationists. Another unlikely pair, Rick Farley, chief executive officer of the National Farmers Federation (not previously renowned for its "green" outlook), and Phillip Toyne, his counterpart at the Australian Conservation Foundation, joined forces and presented Prime Minister Bob Hawke with the idea of ten-year national funding program for Landcare. The Prime Minister adopted the idea. Just four years after Joan Kirner took office, there was nationally supported community action to achieve agricultural sustainability.

In the decade to 2005, more than 400,000 rural community groups have formed across Australia to create local "platforms." (As Niels Röling[31] would say: Landcare is but one, albeit neat, example of a platform of knowledge and communication.) Their purpose is to address agricultural sustainability. They remain, in the main, pretty close to the principles espoused by Kirner and Mitchell:

- Lasting solutions require a whole of community approach, not simply an economic, agricultural, social or environmental response;
- The people affected by the problem or challenge must have their

experiences and expertise valued and be actively part of the solution.
- Effective solutions require government, business, organizations and communities to act in partnership and jointly own the outcomes.
- Departmental services need to be re-structured from hierarchical information providers and policy and funding controllers to empowering facilitators.[32]

And so, how has thinking about sustainability influenced our landscapes? It is tempting to suggest that its impact on agriculture has been minimal, at least in the first twenty years of the paradigm. But this is not so; rather, we take for granted the subtle year-by-year changes in the farmland around us, without reflecting on what has informed the direction of change. For example, in mid-July near Fargo, North Dakota, the landscape is a tapestry of colors: the hills are covered in golden winter wheat, yellow canola, and browning soybeans, and the valley floor is covered in dark-green alfalfa; further along the road near Beach (where "Home on the Range" was written 55 years ago), there are patchworks of wheat fields and grazing land, rotated with legumes (which fix nitrogen from the air in the soil), annuals, and perennials. The mix of legumes, grasses, annuals, and perennials sustains soil carbon and nitrogen and keeps water tables relatively stable.

That being said, the goal of sustainability has had the greatest impact on urban landscapes and city planning. Or, if you believe that urban planning was well underway before the term sustainability was coined, then perhaps it is more honest to say that it has lent a name to the landscaper's purpose. City planners have for a long time seen the value of large-scale planning. This is accepted by citizens, and governments are admonished when they avoid this task. For example, when the Sydney City Council approved plans to build multi-storey apartments along the foreshore of Sydney Harbour up to the Opera House, public protests were sufficient to have the development rescinded. It took several attempts, and about fifteen years, before a plan was both approved and grudgingly accepted by the media and the public. The accepted plan, typical of the trade-offs negotiated within sustainable landscape design, contained a large number of apartments to retain financial viability, but also featured a "gap" in the development to preserve sight-lines from the foreshore to trees in the botanical garden on the headland. It is encouraging when contemporary landscape design, through public pressure, creates multimillion dollar fig trees!

Agricultural landscapes, by contrast, have changed mostly by omission. Crop choices and sequences are now better informed, as the example of the Dakotas illustrates. However, land is ploughed, crops planted, greenhouses built, houses and roads carved through, all on the basis of productivity alone, or in a sequential approach to "triple bottom line" sustainability. This involves devising plans for land use based on economic margins (productivity, or perhaps short-term economic sustainability), followed by some consideration of biophysical constraints and social effects. Some of this sequential or piecemeal approach is historical or cultural: it is easier to think about, and discuss, one part of a problem at a time.

This approach is also shaped in part by scientists' predisposition to develop detailed tools that address very specific aspects of an issue. As an example, many generations of work have been carried out on soil mapping and soil capability classifications. Some of these are one-dimensional, such as identifying the texture of the soil; others are more complex, involving measurement of perhaps six or eight soil attributes. Soil capability categories are used by planners and governments in identifying areas of land that may be used for different types of agriculture or where urbanization may be permitted. Others have developed computer-based crop growth models with varying levels of sophistication and complexity.

Recently, it has become possible to combine relatively few soil attributes, simply measured, with calculated estimates of crop growth and, importantly, to make these calculations for a span of years so that there is some recognition of the variability or risk associated with the resultant indicators or classes. As an example, "land quality" may be estimated and the simulations run according to whether the land is used for conventional cropping or organic farming. One study, based in the Netherlands, in which the results were instructive, illustrates that "quality" or "biophysical sustainability" depends on the purpose to which the land will be used. Here, cereal yields were calculated for fifteen years using real weather data and a large set of commercially relevant rates of application of fertilizer. The calculations were made for two management regimes: conventional tillage (with ploughing, inorganic fertilizers, and pesticide use) and organic farming. The risks in the yield estimates were thus related to yearly variations in weather, the likelihood of unacceptably high leaching of nitrates out of the root zone and into the groundwater (in contravention of the Nitrates Directive), and the impact of the crop management regime on these two outputs.

In the paper that reported this work, the scientists arbitrarily illustrated outputs and risks by choosing to identify the yield that, over the run of fifteen years, had a 20 per cent probability of being exceeded. If society accepts a 10 per cent chance or risk that leaching will occur and contaminate the groundwater, then the land in the case study would be found to be highly suitable for both forms of agriculture, with little to choose between the conventional and the organic management: both had high land quality values of about 86 (on a 100-point scale). If, however, leaching is never allowed, then the land would be seen to be much less suitable for intensive cropping, and a big difference would be perceived between the conventional farm and the farm that is managed organically and has higher levels of organic matter in the soil. Because fertilizer rates have to be lowered in the conventional management to avoid even the possibility of any nitrate leaching, its land quality drops to 33, while under organic management, the value is 61.[33]

The crop production–nitrate example can be played out to explore or predict many other outputs. The value, or fitness, of an area for maintaining bird communities, as another example, may be related simply to the extent of forest cover (for some bird species, anyhow). Here, existing conditions as well as length of forest boundaries and degree of fragmentation may be taken into account in creating an index or measure of sustainability of a particular landscape. Although computer modelling underpins the estimation of crop potential and possible nitrate leaching for a specific field or region, remote sensing using aerial photography, radar, or satellite imagery is also an important tool in the estimation of larger-scale landscape structures, such as the areas of forests and agricultural fields, and of changes that occur over time as land use shifts.

A good approach to calculating the aggregate use of resources, and their wastage, is "ecological footprint" accounting, devised by Mathis Wackernagel and William Rees in 1996 and refined since.[34] This provides a way of estimating the resources we use to maintain our various lifestyles. Its application, of course, is much broader than just agriculture and speaks to the issue of the sustainability of societies and, indeed, of the world's ecosystems.

The ecological footprint is an estimate of the amount of the world's biological productivity that a person, on average, uses in a year for food, housing and other infrastructure, and consumable goods, and to absorb the greenhouse gas emissions that are entailed in the production of inputs such as electricity, aluminum, and plastics. At present, the footprint does not estimate the area of land needed to produce fresh water, nor does it take account of the impact of wastes. In the case of carbon dioxide emissions, the total emissions by a country are divided by the carbon-fixing capacity of forested land to estimate how much forest should be added to create a full footprint for that country. Imports, and what it takes to produce them, are added to a country's footprint of consumption; exports are subtracted.

Some land area uses are well documented. Worldwide, land to accommodate houses, transportation, industries, and to capture hydroelectric power occupies 0.3 billion hectares. Other areas in use have different levels of productivity, and so the technique attempts to standardize these areas of land and ocean to units of common productivity. The world, according to this analysis, has a surface of 51 billion hectares, of which 14.4 billion are land; 9.1 billion hectares of this land are usefully productive. Some countries occupy this useful space so that, issues of global equity aside, they can maintain a lifestyle by using more hectares of productive land than can citizens of other, less well-resourced countries.

In the year 2000, the average citizen of the United States used 9.7 hectares of resources to maintain his or her lifestyle. The United States, more richly endowed with natural resources than most countries, had 5.3 hectares of natural resource productivity (standardized, using the footprint technique) per person, at the current population level. Thus, it is "expropriating," through imports, 4.4 hectares per person of someone else's resources. Europe is also using more than is sustainable within its own boundaries: France, Britain, Spain, and Italy use about 4 to 5 hectares per person. These lower numbers do suggest, though, that the ecological footprint of a society can be halved, at least, while lifestyle is maintained or even enriched (if you prefer the low-car, high-coffee culture of Europe to that based on the suburban supermarket in North America). Not surprisingly, the world's largest populations, in China and India, use 1.5 and 0.8 hectares per person respectively. Similarly, Nigeria uses 1.1 hectares. All three of these countries, like those in the developed world, are using more than their natural resources will support. Among the countries with large populations, only Indonesia and Brazil have lifestyles that use fewer resources than they have within their national boundaries.

However, when equity is taken into account, it is clear that the world has reached, or even exceeded by 20 per cent, its limit of sustainable consumption, calculated by the footprint method to be 1.9 hectares per person. If the population rises, as it will to about 10 billion within 50 years, then either we will mine our resources further in a way that cannot be replenished, or our average footprint will have to be reduced to 1.1 hectare per person. Taking the equity case, it is clear that in the year 2000 lifestyles in developed countries were not sustainable. If everyone's share of ecological productivity is 1.9 hectares, then the United States and half a dozen other developed countries including Australia and Canada were using three to five times their share.

Although it is usually publicized at the country level, the ecological footprint approach can be used to check the resource-efficiency of any product, such as food. Although food production is no longer a big issue for the affluent world relative to other consumption, it remains a primary concern in low-income countries. Countries such as Nepal, Haiti, Chad, and Mali have national ecological footprints of 0.8 to 1.1 hectare per person. It is possible, then, to effectively feed and house a person from 100 square metres of productive land, just as it was in old Europe. Equally, the footprint technique can be used to evaluate the cost in natural resources of a new technology or product. For example, production and use (over a three-year span) of a personal computer uses 0.18 hectares.[35] This is an underestimate, not taking account of recycling costs or the water – about 74,000 litres! – and fuel that are consumed in the production of the machine and the generation of electricity to run it.

The ecological footprint gives two warnings. First, inequities based on resource appropriation by one country from another are getting worse and, depending on your personal ethics, seem to be intolerable at the start of the millennium. Second, it is reasonable to conclude, based on this technique, that human consumption levels, given current technologies, have now reached or exceeded the capacity of the world's natural resources: the capacity of our agricultural lands, forests, and seas are now fully used, if not exceeded. The "ecological deficit" – the amount by which we are over-using our resources – represents "mining": resource degradation that will, in time, widen the gap between an unsustainable lifestyle and a degrading resource base. Figure 63 exemplifies a contemporary approach to sustainability, although its philosophy owes more to a tradition that spans millennia than to the paradigm that, although over twenty years old, has still not narrowed the ecological deficit.

Local and regional concerns that stimulate community action such as Landcare, and national and global studies such as the measurement of ecological footprints are having an impact. We are entering an era when regional plans are genuinely based on the philosophy, or paradigm, of sustainability. For example, in the Niagara Peninsula in Ontario, where the Falls create one of the world's greatest tourist attractions, there are fewer greenhouses than might have been built had the regional municipality not recognized tourism and aesthetics

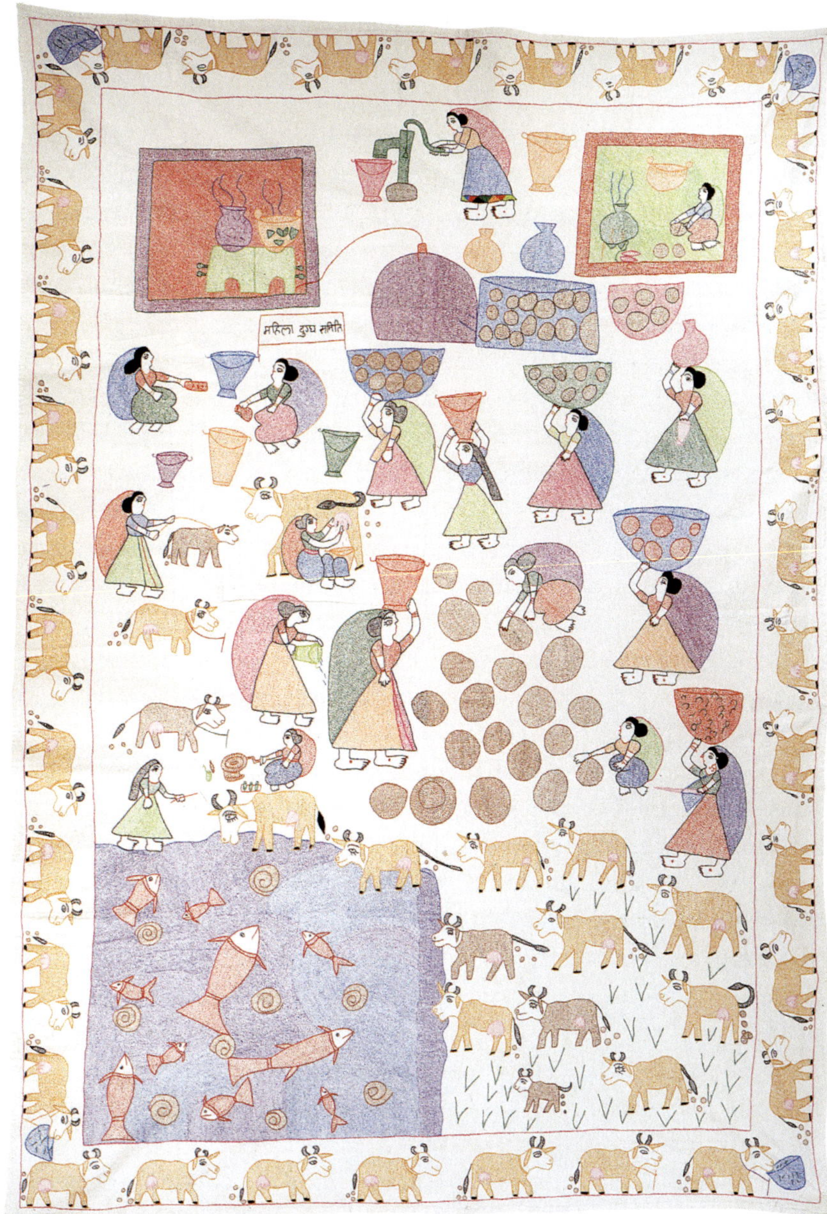

Figure 63

Cow Dung (sujuni textile). Drawing by Nirmila H. Embroidered appliqué by women in Bihar, India. The notion of sustainability and the "small is beautiful" ecological footprint are exemplified in this remarkable appliqué, which depicts the uses of dung, an important economic and agricultural product in this small village. The background is hand-spun, hand-woven cloth made from silk produced by the women themselves. Traditionally, the production of silk cloth was a male activity; however, many farmers have gone to small towns to seek work, leaving the women to take over silk production. The women have also revived their sewing traditions by working for a co-operative that sells their textile works. These sales generate important income for their families. The women have also created AIDS-awareness appliqués as a public health warning; almost everyone in the village is illiterate.[36]

as elements to consider in planning the regional landscape and economy. Viticulture, greenhouse flowers and vegetables, regional malls and fun parlours, falls and parklands, suburbia, electricity generation plants, and freeways to carry people and produce – all must connect and co-exist. Plans for highways are developed simultaneously with plans for regional greenbelts.

Yet the issues remain. Highway 401 is a grey ribbon stretching from Montréal through Toronto to the Windsor/Detroit border crossing into Michigan. At any time of day or night trucks and cars travel too close to one another for comfort at about 120 kilometres per hour. Some of the trucks, clean and slick-sided, advertise the producers of flowers or fruit or mushrooms; others, carrying the brand names of the retail stores they serve, re-ship the same food from distribution complexes larger than the factories of Old Europe to grocery store shelves. It is said that each item of food now travels two thousand kilometres between where it is grown and the house where it is consumed.

Other trucks that travel the 401 are unmarked. Each day they cart five thousand tonnes of Toronto's garbage to Michigan for burial. These trucks, and the agriculture and international food distribution systems they maintain, represent both the opportunity and the challenge of the emerging paradigm of connectivity.

6

MILLENNIUM ACCOUNTING

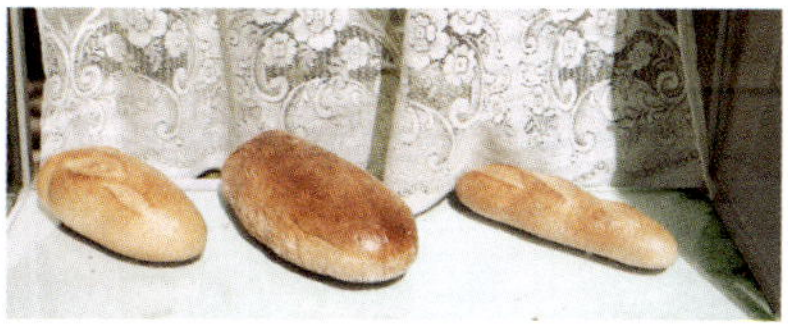

After World War II, when the weary world was fearful about national food security, the image of the "good farmer" persisted, as it had throughout history. By the start of the twenty-first century, however, this image had been replaced with that of the "poor farmer," and even the "bad farmer." To examine the meaning of these judgments is to demonstrate that an accounting of agricultural practice, even at a single point in time, is a daunting task that necessarily depends on personal perspective. But it may still be useful to attempt such an accounting. It is all too easy nowadays to be swamped by information on any particular aspect of agricultural "progress." The simple attractiveness of a one-dimensional outlook often leads to shrill exchanges, as in debates about the benefits of commercializing genetically modified organisms to feed the rising global population, versus their harmful impact on small farmers and the environment. A multi-faceted perspective will necessarily contain ambiguities, but it can lead to a platform of agreement for "going forward."

What could we include in a multi-faceted accounting of Western agriculture at the start of the twenty-first century? Is agriculture meeting the basic condition necessary for civilization: feeding people? What of its use of resources to support new technologies and contemporary lifestyles? Does it value the landscape and natural resources? What, at the start of the century, is the balance between environmental degradation and remediation? In considering whether the world has achieved conditions necessary for further agricultural development, we might ask about government corruption, government regulations relating to the environment and public participation in agricultural and environmental development (recall the indicators that came out of the Davos analysis of indicators of sustainability).

At the beginning of the new millennium the world population stood at about 6.1 billion. It increases at a rate of 210,000 people per day. About the same number – 800 million – are hungry today as were hungry fifty years ago, when the population was only 2.5 billion. How much progress have we made? How many people will go to bed hungry when the population reaches 7.5 billion in 2020?

The global production of food was valued at about US$1.5 trillion in 2000. Outputs continue to rise faster than the increase in population: since 1960, crop food output has doubled, and the production of animal-based foods has tripled (Figs. 64, 65). This has been accomplished with a relatively small increase in the total land area devoted to agriculture (see Appendix 3): although continued expansion has been significant in particular regions such as southwest Australia, Indonesia, and Brazil, the overall pattern has been one of increasing productivity with marginal expansion.

Indeed, in many developed countries the area of land devoted to agriculture declined before 2000. In Canada, for example, the 7 per cent increase in agricultural land that occurred in the west between 1961 and 2001 was more than offset by removal of land from agriculture in the Atlantic Provinces, Quebec and Ontario (see Appendix 2). Viewed from a personal perspective, these changes are dramatic. Years ago, Ontario's Highway 401 ran through intensively cropped fields and dairy farms, even across the "top" of Toronto. Today, "ribbon" development along the highway, together with suburban infill behind it, has covered a further 60 kilometres from the junction of Highway 217 where there was a dairy farm forty years ago, to the edge of the Niagara escarpment, now preserved as a UNESCO World Heritage area.

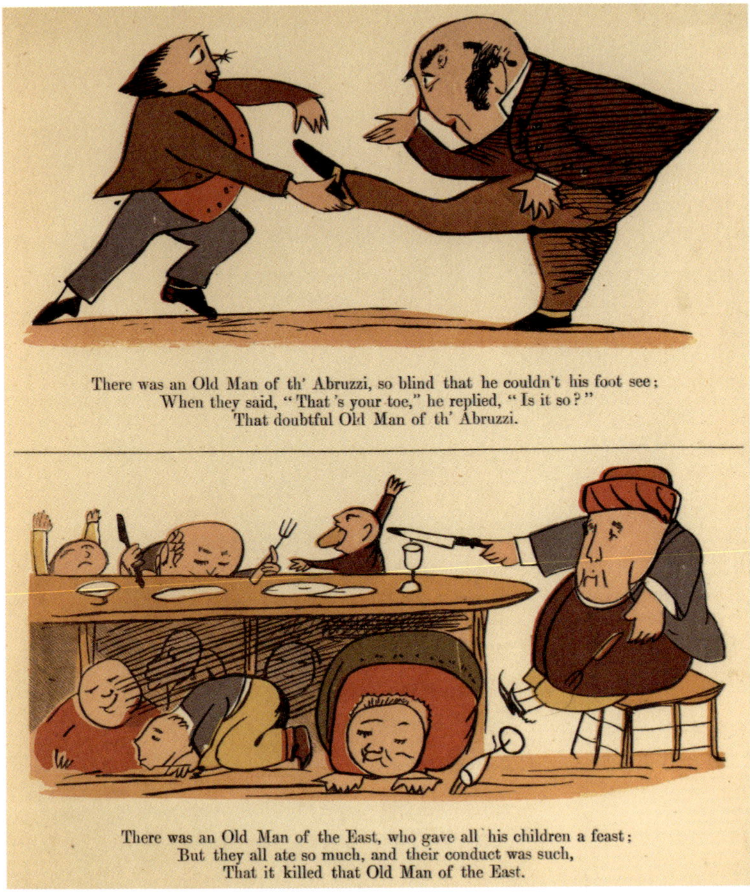

Figure 64
Edward Lear, *The Old Man of the East* (from *A Book of Nonsense*), circa 1866 (silkscreen in colours). Many artists have commented on the developed world's fixation on eating and the tremendous waste of food by the affluent. Obesity was widely acknowledged as a major health hazard in the United States by the end of the millennium. Artists like Edward Lear (1812–1888) recognized the problem a century and a half earlier. In *A Book of Nonsense* he commented on social behaviour through illustrated limericks. His ridiculous rhymes remain popular today: the "old man of the East" is bankrupted (killed) by his children's gluttony.

According to the World Resources Institute, the key measure of agricultural progress – food supply per person – was 24 per cent higher in 2000 than in 1961, while real prices were 40 per cent lower.[1] Further, the freshness of food and its freedom from (or more accurately, tolerably low level of) microbial pathogens are better than ever, at least in the developed world.

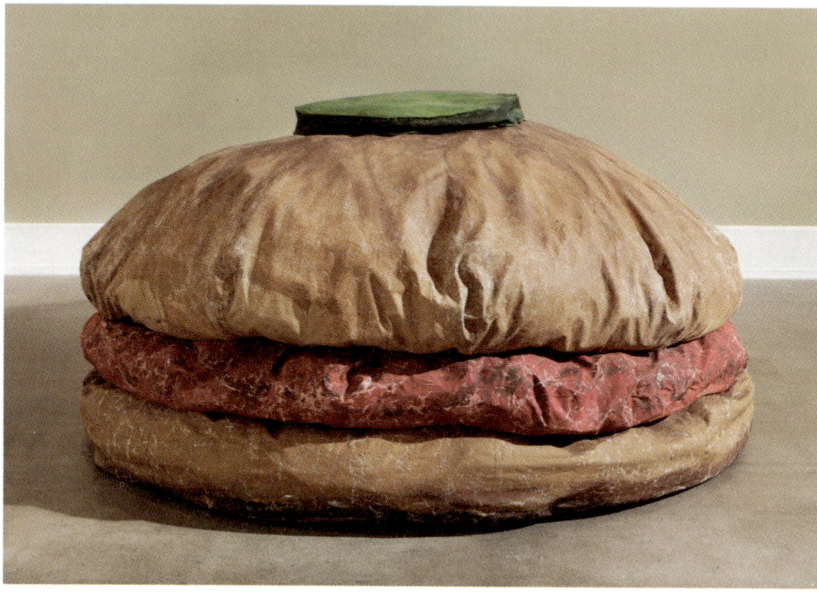

Figure 65
Claes Oldenburg, *Floor Burger*, 1962 (canvas filled with foam rubber and cardboard boxes, acrylic paint). Oldenburg's famous giant hamburger, complete with a pickle on top, is a monument to fast-food culture. By the end of the millennium, a large proportion of North American families were increasingly relying on the high-fat, high-sugar and high-salt combos of fries and burgers and the like for their dietary needs. The convenience of fast food and irregular eating habits is, for many, replacing the traditional family dinner of meat, potatoes, and two servings of vegetables.

Driven by science and technology, market competitiveness, and purchasing power, food quality is high. Science has, over twenty years, embedded food risk analysis in our food processing and delivery practices, largely below the awareness of consumers. "Hazard analysis and critical control points" (HACCPs) are standard within firms, reflecting government regulation and the practice of random monitoring. Unprecedentedly cheap transportation has increasingly allowed retailers to source high-quality food from anywhere in the world. In the large supermarkets of Europe and North America, the seasons have almost disappeared as we purchase fruit and vegetables from both hemispheres, year-round.

The other driver, purchasing power, also causes quality to continually improve. In the United States and Canada our annual food bill, including money spent at restaurants and fast-food outlets, has for the average person been earned by mid-February, and the amount of money that is returned to the farmer has been paid for a month earlier. This means that affluent consumers have a lot of buying power, and discretion, to offer extra money for higher quality food. The growing demand, especially in Europe, for locally and organically produced food (which may or may not be of higher quality) exemplifies the power of this. There is also an increasing demand for diversity in fruits and vegetables, particularly those that are flavourful. The "good taste movement" and "slow food movement" encourage small growers to revive older varieties to meet the needs of restaurants and specialty markets. One high-end Toronto grocer, for instance, offers twenty-three different kinds of tomatoes.

More generally, at the beginning of this twenty-first century there is a growing willingness to pay for foods that are relatively unprocessed. Often, freshness is bundled with a price premium for local, freshly harvested food, as in the case of North Americans who buy most of their food from an air-conditioned supermarket and then drive out of town to a farm shed to purchase freshly picked corn. At other times, a premium is paid for efficient delivery of fresh produce from Guatemala or Honduras.

Today, safety is improved continually through commercial self-interest and government agencies, which in combination have created regulatory requirements for the labelling of food and surveillance systems that cause specific food items to be recalled if they are found to be contaminated, or have deteriorated, or are otherwise unsafe. In some countries, packaged foods are stamped with an expiry date before which they must either be sold or withdrawn from store shelves. The United States, in free-enterprise fashion, does not have uniform "use by" date labelling (although the dating of dairy products has been practised since 1917). Britain, after publicly debating the merits of labelling to improve food safety for fifty years, recommended sell-by date labelling in 1976 and enforces the European Community's uniform laws of 1993. In a similar vein, as the millennium approached, Australia and New Zealand joined forces to create a single regulatory agency, now called Food Standards Australia New Zealand.

In Europe, in particular, there has been a strong negative reaction to the sale of genetically modified (GM) foods, whether they are appropriately labelled or not (Fig. 66). This ethical and health concern is also compounded by the public's anxiety about the possible damaging result of uncontrolled scientific experimentation. The issue of food labelling to allow consumers the choice of whether to select GM foods has been ongoing since the mid-1990s and has still not been universally resolved. By 2000 there was a spectrum of approaches. In 1997 and 1998, two supermarkets, Iceland and Carrefour, introduced bans on GM foods in their stores in the United Kingdom; these were subsequently withdrawn, and in 1999 eight supermarkets (representing 90 per cent of retail sales) introduced a voluntary policy of requiring labelling of all GM-derived foods. From 2001, Food Standards Australia New Zealand required that if a final food contained novel DNA or a novel protein produced by means of genetic technology, then the food should be labelled as genetically modified. Exemptions are for flavourings and highly refined foods such as sugars and oils, in which the processing causes the genetically modified DNA or protein to be denatured or removed. Community expectations vary: some countries, such as the United States, have announced that they will not require labelling for foods containing genetically modified substances, while community attitudes are still evolving.

Thus, at the beginning of the millennium large multinational retailers in Europe are labelling or excluding GM-derived foods in the expectation that this will give them better market acceptance, while elsewhere GM-derived foods permeate the shopping experience. In 1998, when Carrefour drew attention to their GM-derived foods, they identified genetically modified organisms in 516 out of 1,800 own-label lines. It is likely that a "second generation" of GM-derived foods will soon be created with targeted nutritional or pharmaceutical benefits. These will appeal particularly to aging populations in affluent, developed countries where there is discretionary spending-power to pay a premium for these foods. Then, presumably, retail marketing will advocate their consumption, with labelling to allow consumers to exercise choice.

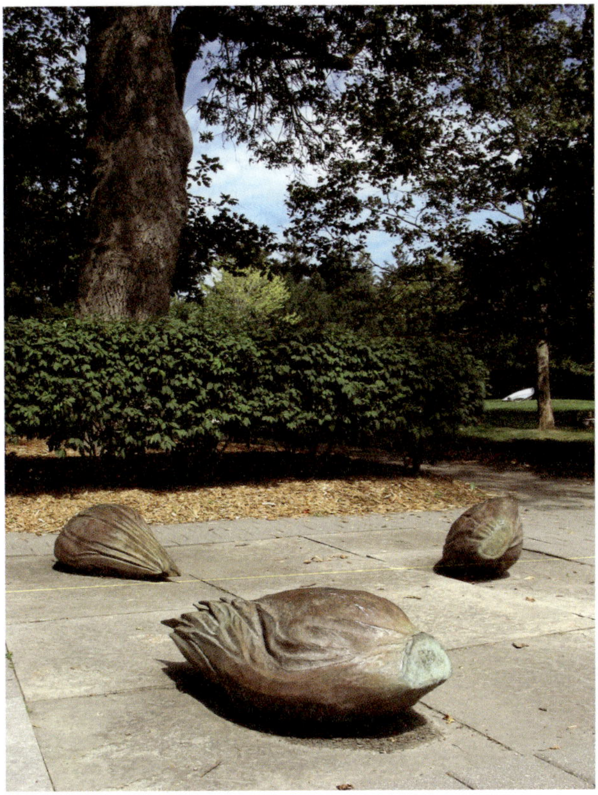

Figure 66
John Greer, *Three Grains of Wheat*, 1991 (bronze). Greer responds to concerns about genetically modified foods. These giant seeds, almost one metre in length, express an amazingly bountiful harvest. They could also signify a biogenetic experiment gone wrong – a Frankensteinian world in which we are enveloped by our own food.

More generally, there is still some ambivalence as to whether labelling benefits the public, or the producer. This doubt underpins ongoing disagreement between consumer groups and manufacturers about the need for labels that give details about the composition of processed foods, whether this relates to genetic modification or other issues such as the source of the food. (The costs of monitoring to ensure compliance is a further point of contention.) In 2000, Ontario wine could be labelled "VQA" (Vintners Quality Alliance) only if it contained 100 per cent Niagara grapes, and "Ontario" if it contained 30 per cent Niagara grapes; after two disastrous harvests the industry successfully petitioned for a change in labelling so that, in 2005, Ontario-labelled wine could contain as little as 1 per cent Niagara grapes: hardly helpful to the consumer, who presumably tastes the other 99 per cent within the bottle!

As food on the market shelves approaches its use-by date, it is heavily discounted to encourage quick consumption, or gathered up by the retailer and redistributed to charitable institutions, or thrown out. Recalls of unsafe food are mandated by law in most developed countries. As surveillance systems improve, the number of recalls rises. In the case of age-related spoilage, which commonly account for about one-third of withdrawals from sale or safety warnings, unsafe levels of bacteria and other microorganisms build up in the food; contamination from pesticides, drug residues, and other chemicals accounts for an equal or larger proportion of food recalls.

One of the food safety agencies with quite comprehensive powers is the Canadian Food Inspection Agency (CFIA), established in 1997 to "enforce" food safety and nutritional standards relating to no less than four government portfolios: agriculture, food, health, and industry. The high stakes and impact of these agencies is illustrated by the response of CFIA and others to the finding that one cow in Alberta had mad-cow disease (bovine spongiform encephalopathy) in 2003. The export of live animals and meat to the United States was immediately stopped: a significant measure, given that the annual trade of about 1.7 million cattle is worth C$4 billion. Within 20 days, the cow's herd had been slaughtered, although subsequently none of the animals tested positive for the disease. All cattle related to the infected animal or that might have eaten feed from the same lot six years previously were killed: in all, 1,000 cattle in three provinces. There was no harm done by the diseased animal – its meat had not been sold or consumed – but this unequivocal response illustrates our determination to ensure, and to be seen to ensure, food safety.

Internationally, the Codex Alimentarius Commission, set up in 1963, brings countries together for the ongoing development of internationally agreed food standards. Increasingly, these standards are used to regulate international food trade, including how countries trade with, and how food is treated within, the developing world.

The use of resources to support new technologies and modern lifestyles was dramatically polarized by 2000. The popular press regularly features articles on the different levels of consumption of resources, such as gasoline or electricity, between the United States, other developed economies, and the rest of the world. Nonetheless, public opinion becomes only periodically engaged in the issue of equitable resource use, and the administration of George W. Bush made it clear at the beginning of the twenty-first century that the maintenance of domestic lifestyle in the United States was a higher priority than equity with underdeveloped countries or the impact of consumerist lifestyles on, for example, global climate change.

"Ecological footprint" accounting, devised by Wackernagel and Rees in 1996,[2] gives us a handle on our use of resources, both in agriculture and food production and more generally. High-input agriculture – productivity to excess – provides the basis for cheap, safe food in high-income societies. Expenditure on non-food purchases – the money spent by a household in North America beyond the annual food bill – affects our environment, our natural resources, our capacity to produce food, and our landscapes.

Middle America in the year 2000 can be typified by a 300 square metre, grey and red brick bungalow with white painted trim, standing on a bright green patch of lawn with two or three cars parked between the asphalt road and the garage door. It looks across the road to a house that is much the same, also set square on its piece of private parkland. The road is wide, with curbs and sidewalks; there is a playground at the end of the street and, some kilometres away, a supermarket complex surrounded by a flat landscape of cars occupying what the midwestern farmers would call a quarter- or half-section (remembering that the land was carved, for farming, into sections of one square mile). In Europe, more densely urbanized at the beginning of the new millennium, the average family might live in a whitewashed apartment accessed through a communal balcony and overlooking a walled courtyard. In Asia, Africa, or South America, there is no "average": a family of four or five, together with a mother-in-law or two, might occupy a mansion, a four-roomed house or apartment opening onto the sidewalk, or a galvanized iron and cardboard shack depending on whether they are wealthy, middle income, or urban poor.

When inputs to agriculture are used at a level that is more than necessary to maintain the system, they turn into waste. We can think of inputs to the food chain in three categories: (1) nutrients into the land; (2) energy to drive the various production and transformation systems, such as tractors, canning machines, long-distance road transporters, garbage-removal trucks, and consolidators; and (3) the infrastructure of packaging: the machines necessary to produce the packages and the amount of material that is bound up, inefficiently, in the waste products (Fig. 67). The three-sided challenge that was recognized by 2000, but not yet acted on by society, was to reduce international inequities in resource use, to devise more sustainable (less polluting, lower-input) systems for agriculture, and to create price and reward mechanisms for farmers so they become equitably paid and able to maintain or enhance the natural resource base of their farms.

Despite growing awareness of the wastefulness of excessive fertilizer application, and the harm this does to waterways and wildlife, affluent societies are still coming to terms with the magnitude of the problem and how to deal with it. A particularly stark example is the balance of introduced nutrients in the Murray-Darling river system in Australia. The catchment area of this waterway is huge – about 30 per cent of the continent – yet it contains few cities of any size: the big centres of population cling elsewhere to the coast and drain their wastes directly into sea. In the Murray-Darling Basin, huge inland plains drain slowly toward the sea near Adelaide, although most of the water evaporates on the journey. The plains are so flat, and the river such a vein-like network, that in the north it is called channel country; after the rains, the drainage courses spread out to create sheets of water over thousands of kilometres. Mostly, though, it is an arid basin of contrasts: a narrow river shadowed by grey-green peppermint trees journeying across a plain dazzling with heat haze, covered by saltbush, low eucalyptus scrub, and wheat fields.

There are diffuse as well as point sources of nutrients into the river. Diffuse inputs, or pollutants, arise from the extensive fertilization of crops and waste from livestock over large areas. Point-source pollution arises from hog farms, beef feedlots and population centres,

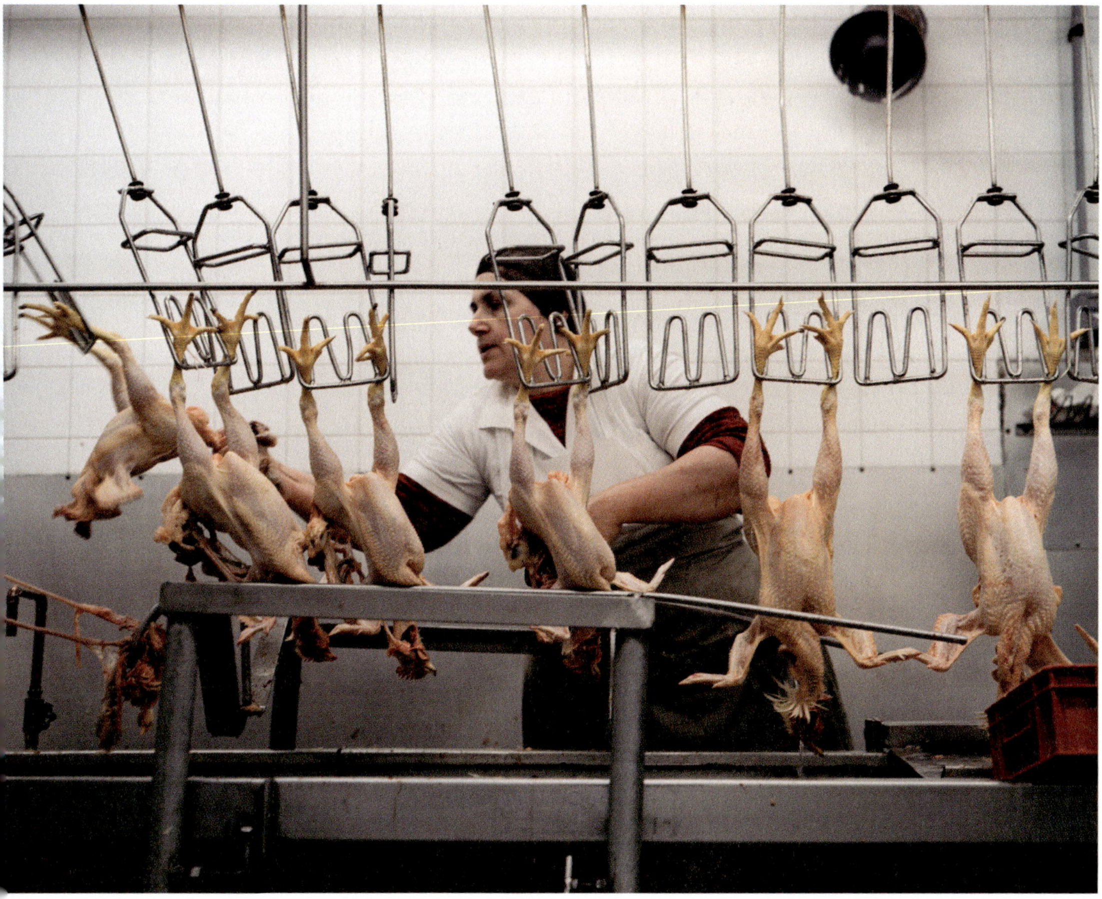

Figure 67
Edward Burtynsky, *Chicken Packaging Plant #1, Toronto*, 1983 (chromogenic colour print). Burtynsky's image of a chicken processing plant is an almost poetic interpretation to the rhythm of the work. Unfortunately, the end product is not only a fine chicken dinner but also a mound of Styrofoam and plastic wrap.

and from agricultural activities that are easy to delineate geographically. The fact that the numbers are large is disturbing, given that this is, although vast, a mostly semi-arid region of extensive (not intensive) farming and of low-capacity rangeland grazing. For example, consultants Gutteridge, Haskins & Davey estimated in 1992[3] that animal feed lots, comprising about 180,000 cattle in the northern tributaries of the Darling River, add about 7,200 tonnes of nitrogen effluent to the river each year, and that over 1 million hogs in hog farms add about 9,000 tonnes of nitrogen. Human sewage, from all the cities and small towns in the Basin, create about half this waste. By contrast, the nitrogen flow into the catchment and the water from diffuse sources, such as irrigation and drainage after fertilization of croplands, might contribute a whopping 27,000 tonnes of nitrogen in a wet year. These are nutrients that are being wasted by "bad farmers" in an affluent society, to the detriment of the environment, while elsewhere in the world, in places such as west Africa, crops produce poor yields for want of nutrients, and people go hungry.

Efficient and less wasteful use of inputs to agriculture, efficient heating, new forms of energy, drastically reduced packaging, local recycling, new building products made from bio-resources (particularly, plants), and that old standby, public transport, will all markedly reduce our ecological footprint and the rape of the landscape. Agriculture will also have to make innovations – and in this respect it has a long track record. Some innovations, such as industrial livestock barns, are heavy users of natural resources (e.g., electricity, transport), although they drive down local (enterprise) costs of production and prices to the consumer. Others provide the way for the future.

Innovation makes sense only if it is financially competitive with respect to environmental inputs or costs and is calculated on the life cycle of the product. Life-cycle analysis involves working out the full sequence of technologies and costs from cradle to grave or, in the increasingly common case of recycling, partial resurrection. The European Union Eco-label award scheme was a forerunner of many schemes to reward life-cycle costings. Established in 1992, its goal was to "promote the design, production, marketing and use of products which have a reduced environmental impact during their entire life cycle."[4] The Eco-label scheme required ecological criteria for impact in all life stages (pre-production, production, distribution, sale, use, and disposal) for each product group, whether copying paper, dishwashers, or T-shirts. Sometimes this accreditation conferred a market advantage; at other times it wallowed in bureaucracy. Most of these schemes were only partly successful by 2000. However, in the food industry the time of eco-labelling may be about to come: promoted by individual retail chains, it is likely to gain wide acceptance and create market advantage, as we will suggest in the next chapter.

How has demand for high-quality, inexpensive food (Fig. 68) and the valuing of natural resources affected our landscapes? There is no simple answer: the impact has ranged from richness and fierce custodianship to degradation that is accepted and sometimes almost celebrated. In Europe, where cultivation is as old as in most regions of the world, one may travel past exposed terra-cotta fields and eroded hillsides in the south of Italy, to irregular-shaped fields under a humid, polluted grey-brown sky near Milan. In early July, stressed corn with upturned droughty leaves is at the tasselling stage, growing up to the edges of roads and abandoned farm sheds. Every 20 kilometres or so is a patch of ploughed soil or a rare, unused field with green grass and wildflowers. Sometimes, Holstein milking cows stand in bare-ground pens, ready for the next milking. Nearer Verona, corn is mixed with vineyards. Further north, as the climate gets cooler, the corn disappears and the landscape is filled with vines on T-shaped trellises designed to intercept the sun; elsewhere are densely planted pears with yellow-green hanging fruit. As sharp as a knife, the Dolomites rise in the distance, a backdrop for the grapes, pears, and apples: sheer cliffs, quarries and brush, contrast with the fruit which is now covered with netting for protection against evening thunderstorms and in the hope of achieving a market premium for quality. Near the Italian–Austrian border, townspeople come off trains in orderly clumps and lines to purchase fresh ripe apricots, a delicacy that will dominate dessert menus through middle Europe

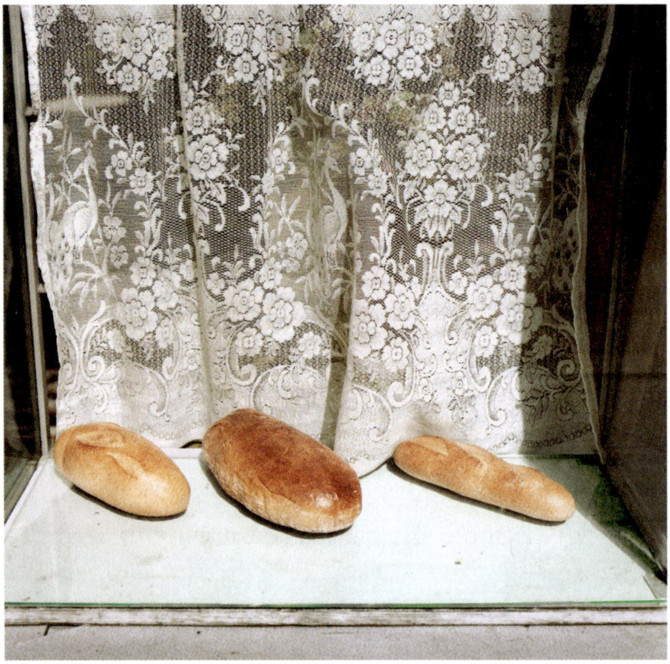

Figure 68
David Hlynsky, *Window, 3 Loaves of Bread, Poland*, 1988 (colour photograph). A baker has displayed his artisan loaves in his shop window, as if he were a curator placing sculptures in a museum case. The emphasis is on good taste and well-being rather than supermarket excess.

for a month. Into the alpine regions and the Rhine and Danube valleys, the landscapes are first meadows with white and pink flowers and remarkably few cattle, then tapestries of wheatfields and vineyards and, where it is warmer, corn again.

The year 2000 is a significant marker in an accounting of agricultural innovation; the turn of the millennium was critical in the stepping-forward of biotechnology, particularly with respect to the creation and use of transgenic organisms. Time will tell, but this will probably be regarded as one of the four most significant technical innovations in agriculture since the social and technical upheaval that transformed the manorial system of old Europe. The other innovations are: the breeding of short-stalked high-yielding cereals;

the discovery and use of fertilizers and pesticides; and the invention of farm machinery. In the five years before 2000 there was also significant development of "precision agriculture" – the use of remote sensing, sometimes by satellite or based on soil testing, coupled with global positioning systems mounted in tractors and spray-rigs – to continuously fine-tune inputs to cropping so that they reflect the resources of the local area and its yield potential. However, as exciting as precision agriculture may be, it can be regarded as an evolution that captures the power of information technology, not a revolution in itself.

Biotechnology owes its origins to work on genetics and inheritance by Mendel, Watson, Crick and others, and to the push by science and industry to accelerate the speed of breeding so that new plants and animals reach the market faster. Biotechnology is an umbrella-word that includes five spheres of activity in agriculture and food. These are:

- Traditional breeding. This practice usually hybridized different varieties of the same species of plant or animal, although there are examples of successful interspecific hybridization, such as the production of triticale (wheat crossed with rye) and the ass (donkey and horse). The conservatism, and reasons for slow progress, in traditional breeding were two-fold: it was not done at the cellular (or molecular) level, so that all the genes of both parents (generally) were mixed randomly to produce a range of all possible offspring, and wide crosses between different organisms, for example, between a plant and a bacterium, were highly unlikely to be stable or survive.
- Bio-informatics. This begins with the sequencing of the genetic code that makes an organism what it is, whether human being or rice plant. Bio-informatics allows us to pin-point DNA sequences associated with particular traits such as colour and disease susceptibility.
- Molecular breeding. Biotechnology allows us to remove small sequences of DNA from one organism and insert them in another, carrying across (if the organisms survive) the associated characteristic, such as a specific disease-resistance, to the new host. This can

be done within the same species, in which case it is like traditional breeding but without the randomness that is unavoidable in whole-plant or whole-animal breeding. It is also done by combining genetic material from unrelated species, for example, inserting a piece of bacterial DNA into a plant. The precision and speed of this technology overcome the constraints of traditional breeding and create a much wider range of possible hybrids.

- Regeneration technology. This involves cloning, taking cells or pieces of tissue and growing them in artificial environments so that very large numbers – hundreds of millions – of near-identical plants can be produced, disease-free, faster than ever before.
- Multiplication technology. Industrial food biotechnology, particularly, uses precisely controlled artificial environments (vats) to multiply particular organisms and then kill them in the production of quick-growing, homogeneous and predictable foods, such as cheese and yoghurt. In some cases these processes may include organisms, such as yeasts, created through same-species molecular breeding; in other cases, they are simply incubators for growth that would occur naturally, albeit more slowly.

Whether society accepts any or all of these activities in mainstream agriculture will be determined on the basis of issues that society perceives from ethical, social, and economic standpoints rather than a strictly scientific perspective (Figs. 69, 70). The issues are not simply transparent, either. For example, in 2000 there was public concern, stimulated by non-governmental organizations, about the export of genetically modified grain to developing countries. The biotechnology might (or might not) have aroused social objections in resource-poor African countries, given that the crops had been extensively trialled, and the grain eaten, in North America. However, African governments rightly asked why they would accept genetically modified grain while its use was being debated in the source country, and while its sowing might compel the subsistence African farmer to return to the seed company for fresh seed each year, and to particular pesticide companies for products to control weeds in the new crop.

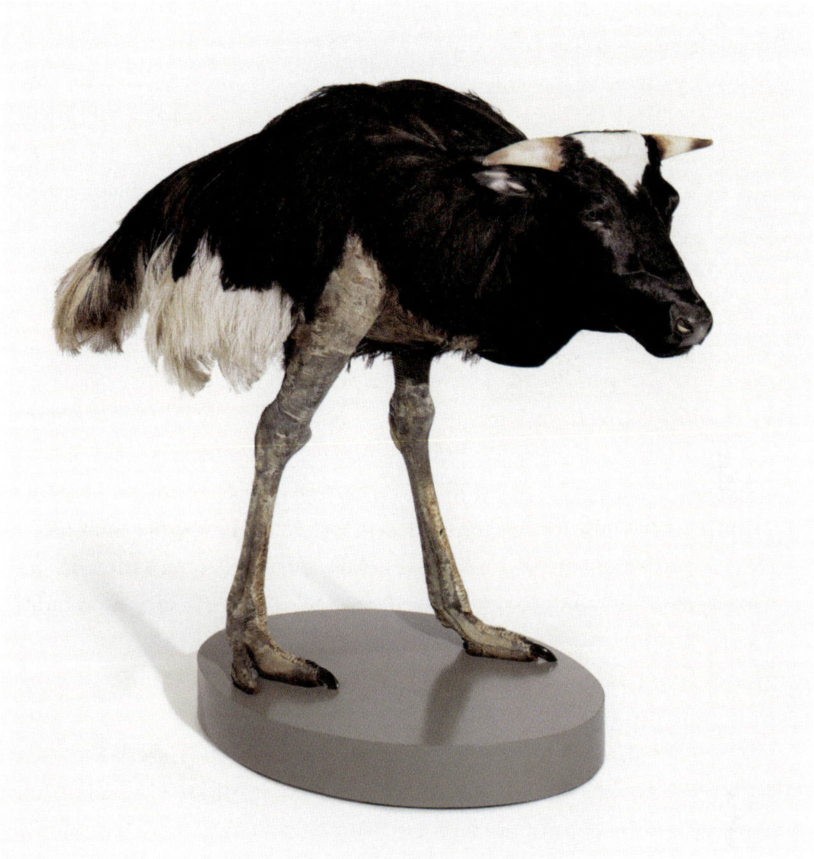

Figure 69
Thomas Grünfeld, *misfit (cow/ostrich)*, 1997 (taxidermy). Grünfeld projects a Frankensteinian world with a series of sculptures called *misfits*. In this example a cow's head is attached to the body of an ostrich.

In 2000, 25 per cent of corn, 54 per cent of soybeans, and 61 per cent of cotton grown in the United States were identified as genetically modified. These numbers rose to 34, 75 and 71 per cent respectively by 2002.[5] Pollen flow between the GM crops and conventionally bred crops means that, in many cases, seed is hopelessly intermixed: we have lost the capacity to separate GM from non-GM canola, for example. The GM wheat that is called "Roundup-ready" in reference to its

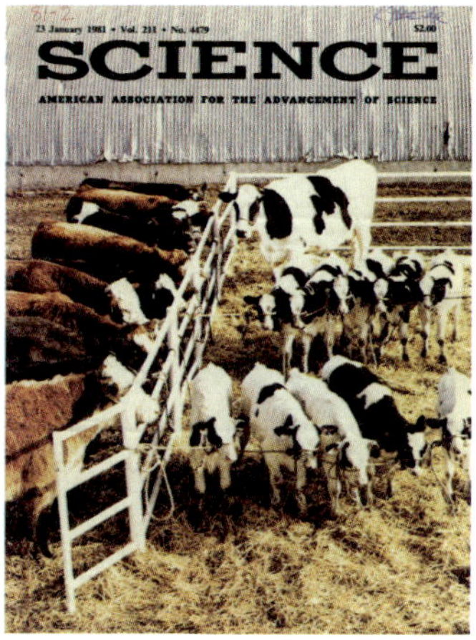

Figure 70
This cover from a January 1981 issue of the journal *Science: American Association for the Advancement of Science* shows an image of a dairy cow that has borne calves through embryonic transfer. The legend reads: "The dairy cow (upper right) is the genetic mother of the ten calves. She was superovulated, and the embryos were recovered from her uterus 1 week after conception." Similar technology, also not involving genetic modification or transgenesis, is nowadays applied to women. Embryo transfer is used to improve the quality of the animal product and overall reproduction. More than half a million bovine embryos were transferred in 2003.[6]

capacity to withstand the blanket spraying of Monsanto's herbicide Roundup, will, like its canola cousin, create significant problems of weed control. Although a GM crop makes weed control simple and, some argue, more environmentally friendly during the life of the crop itself, the herbicide-resistant "volunteers" – GM seedlings that grow in the next season from grain spilled in the field – will add cost and complexity to weed control in later years.

Agricultural innovation will depend on the recognition that agriculture will play a central role in creating the coming "bio-resources economy." This economy will be based, among other things, on heavy use of agriculturally produced renewable resources. Consider corn. Corn used to be harvested and turned into polenta and breakfast cereal or eaten fresh on the cob. It still is; however, by the millennium, processed corn was being used in a wide range of products that, depending on price, could substitute for petroleum-based or petroleum-dependent goods. For example, cake mixes sometimes use pre-gelatinized corn starch, which forms a paste in warm or cold water; corn cobs, when finely ground, are very absorbent and so can be used as carriers for cosmetic creams, vitamins, soaps, and as absorbent animal litter. Rather than using petroleum-based solvents, corn can be processed to create a resin, tetrahydrofurfuryl alcohol, which is used as a solvent for dyes, resins, and lacquers in the paint industry. Similarly, soybean grain, which has an high oil content, is used as a partial substitute for petroleum to fuel cars and tractors.

Perhaps agriculturally produced fuels would be dominant today if Henry Ford had not decided it would be easier to use petroleum for his cars rather than grain-based alcohol, which was a real alternative at the beginning of the last century.

Although the bio-resource, renewable economy contributed only a minor share of the whole economy in 2000, it will grow as we increasingly use common sense to value the natural environment and the inputs we take from it. The field of "environmental economics" has grown to address this; the term dates from the 1980s, but this approach did not begin to seriously rival neo-classical economic analysis until the beginning of the millennium. The ecological footprint, as we have discussed, is a complementary approach to valuing ecological services, although it uses area, rather than money, as a unit of measurement and values only things that are produced rather than an array of both tangible and intangible services.

A quotation from 2000 exemplifies the new attention that the paradigm of sustainability brought to the valuing of "ecological services" at about this time:

> Conventional indicators of economic growth grow progressively more misleading as market and policy failures, and the inefficiencies they induce, multiply. Fluctuations in economic growth and the distribution of its gains can lead to alternating regimes in which resource degradation is driven first by poverty, then by growth, and then by poverty again.[7]

A figure to reflect on: ecosystem services such as landscapes that preserve and create clean water, and insects that pollinate our food plants, are estimated to contribute US$33 trillion to the total global economy. This is about twice the value of human-made things – our global gross domestic product of US$18 trillion, measured regularly through conventional economics.

Ecosystem services maintain biodiversity and the production of ecosystem goods, such as food, grass for domestic livestock and wildlife, timber, biomass fuels, natural fibres, many pharmaceuticals, and industrial products. The harvest and trade of these goods represent important and familiar parts of the human economy. By 2000 it was recognized that, in addition to their role in the production of goods, ecosystem services support life through, for example: purification of air and water; mitigation of droughts and floods; generation and preservation of soils and renewal of their fertility; detoxification and decomposition of wastes; pollination of crops and natural vegetation; dispersal of seeds; cycling and movement of nutrients; control of the vast majority of potential agricultural pests; maintenance of biodiversity; protection of coastal shores from erosion by waves; protection from the sun's harmful ultraviolet rays; partial stabilization of climate; moderation of weather extremes and their effects; and provision of aesthetic beauty and intellectual stimulation that lift the human spirit.[8] It is a long, but incomplete, list.

As an outcome from growing commitment to sustainability, a few governments have commissioned "state of the nation" studies. What have these studies told us about the agricultural environment in 2000? Broadly, the results were disappointing. The World Resources Institute said in 2000[9] that 9 per cent of the world's agricultural land was "strongly or extremely" degraded and that 43 per cent was moderately degraded. Their "agroecosystem scorecard" includes statements such as "salinization is estimated to reduce farm income by $11 billion each year" and "fully 70 per cent of water withdrawn from freshwater systems" is used for irrigation, and only 30–60 per cent is returned for downstream use. The aggregate figures are staggering: water erosion affects a 1.1 billion hectares, and wind erosion half of this. The rate of erosion – soil shifted, largely as a result of human activities – is in the order of 1.9 billion hectares. Overgrazing, deforestation, and agriculture contribute most to this loss of soil.[10]

The millennium landscape of the "New Worlds" varies of course, depending on region. Dry croplands, such as those in Saskatchewan and Western Australia, have not fared well. Flying over the spare, almost emaciated "great southern" wheat, cattle, and sheep farms of southwest Australia brings to mind a remark made by the painter Fred Williams shortly before he died: "My God, from here it looks just like I painted it." Williams's skeletal landscapes are the outcome of how farmers created, or at least modified, the landscape (Fig. 71). Remnant patches of native vegetation – scrub – remain, and big swathes in national parks along marginal lands, such as stony, sloping terrain, or land where the early settlers could not find surface water for their livestock and their homesteads. The cropland in late spring looks dry: golden, but crackling with dry air and dust, and as the plants die and the golden leaves fall, more of the yellow or grey sand is visible. It is different here in autumn and winter, when apple-green crops and the moist soil lend freshness and optimism. In Canada, of course, it is the early summer that is fresh and green, with the wheat golden in autumn and the prairies white, desolate, and dangerous through the winter (Fig. 72).

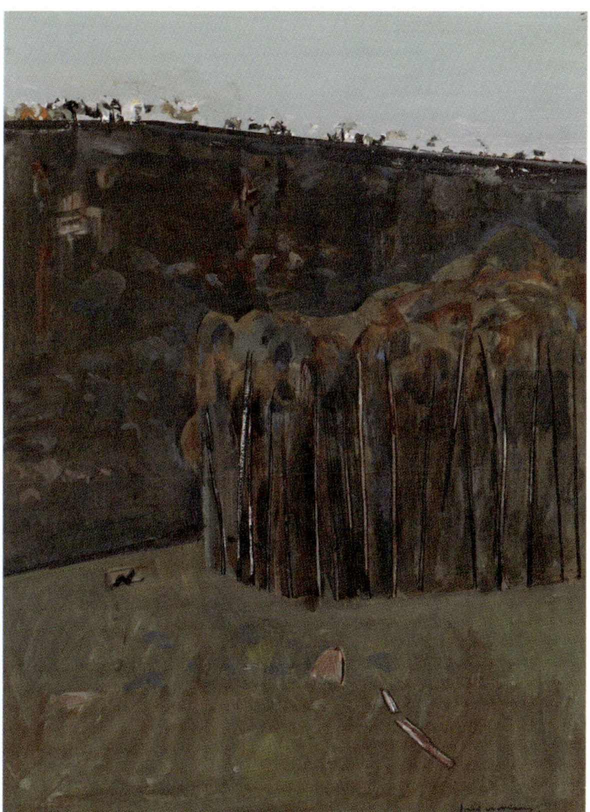

Figure 71
Fred Williams, *Saplings, Mittagong*, 1961–62 (gouache).

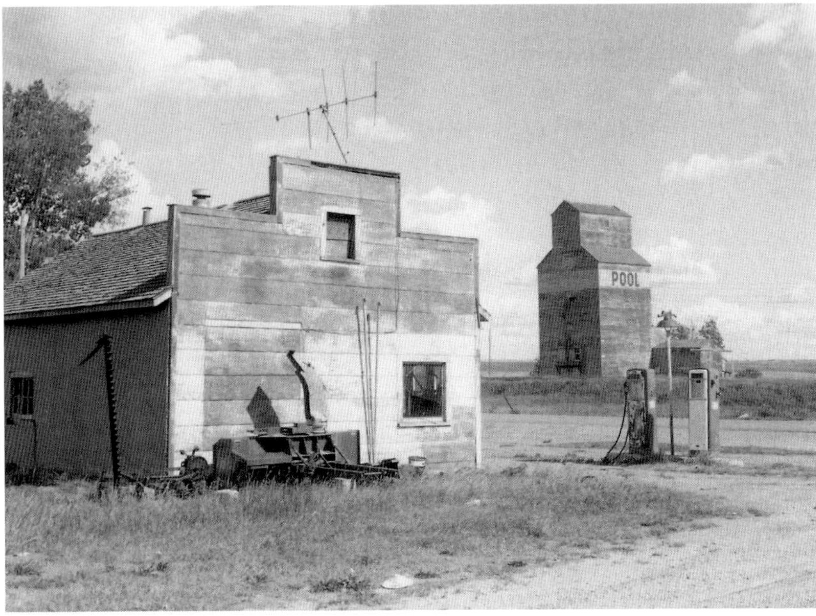

Figure 72
Orest Semchishen, *Springwater, Saskatchewan*, 1983 (colour photograph).

The croplands contain towns where, thirty years ago, shops and pubs lined both sides of the road: quiet places even then, when lunch was a choice of a pie from the general store or a pie on a plate and a beer at the pub. Today the croplands are dying, almost dead, or thriving, and communities have become polarized accordingly. As agriculture has come to support fewer services, people have moved away to create bigger towns or small cities, where they are able to access doctors and hospitals, and get a daily newspaper. The small towns dry up, the shop-owners retire, and the newspapers are posted over empty windows. The dying speeds up when the local medical service or nursing station, and then the bank, closes; then it is only a matter of time before the gas station and the operator of the small "independent" supermarket realizes that customers are picking up their groceries when they travel 80 kilometres each way for the once-a-week bank, shop, and social (Fig. 73).

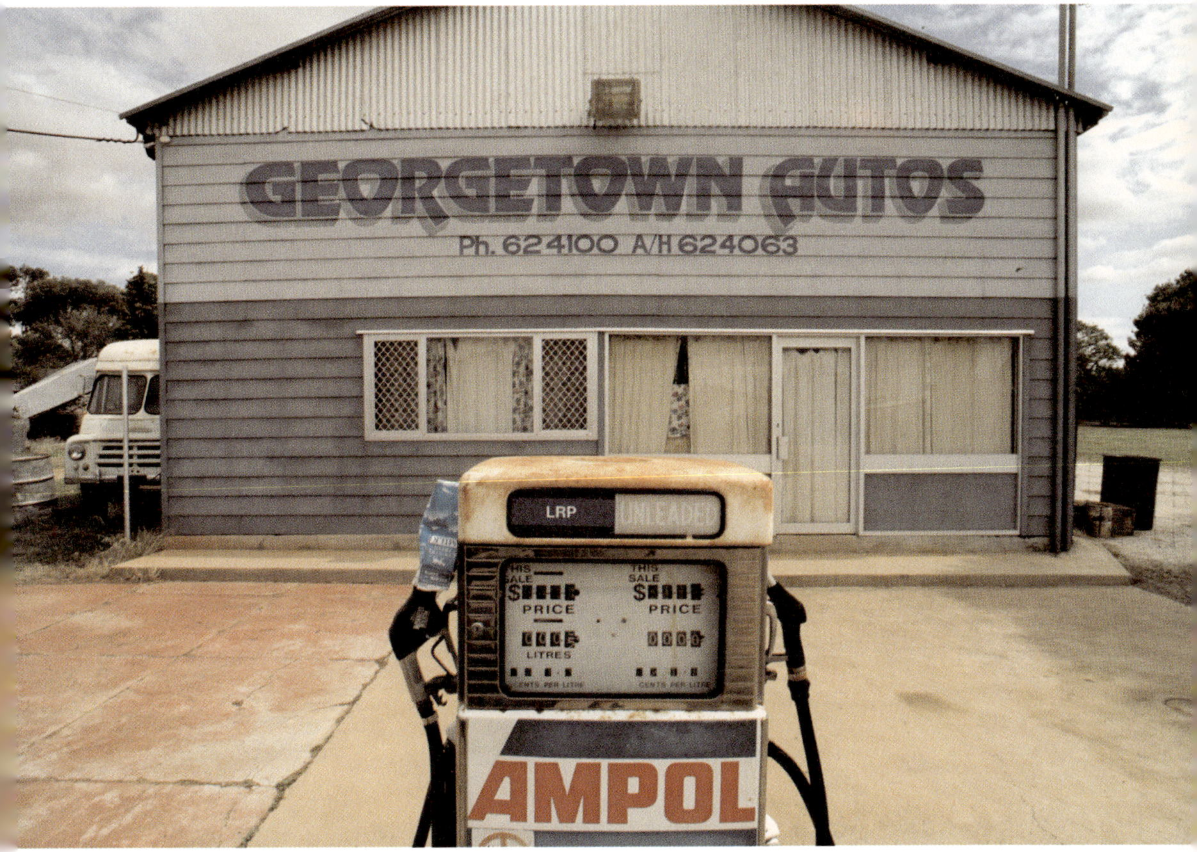

Figure 73
Tomek Sikora, *Georgetown Autos*, 2003 (transfer on card). Polish artist Tomek Sikora created this photograph of an abandoned gas station in an outback town in Far North Queensland, Australia. Georgetown was founded in the 1870s as a gemfield mining town, and since 1900, it has been primary grazing country; however, the town is showing signs of decline. Just as Sikora's photograph addresses the perilous state of farming in Australia's rural locales, Canadian journalist Ingeborg Boyens's book *Another Season's Promise: Hope and Despair in Canada's Farm Country* (2002)[11] captures a similar sense of hopelessness. Her focus is the impact of industrial agriculture on individual farmers in grain-farming communities in Saskatchewan and their ability to meet demands for production efficiency and quantity.

Rural landscapes from Saskatchewan to South Australia have become depopulated. Settlers cut farms out of eucalyptus forest in the "high country" of New South Wales, Australia, where the balladists wrote "the air is clear as crystal." As recently as the 1950s, the local village – which was some distance from the highway – consisted of a general store, fruiterer, one-room schoolhouse, and five or six homes. The general store could have been from a thousand rural villages: two steps up, a creaky wood-framed fly-screen door, a bare-boards floor and an on-site liquor licence, the back of the room serving as a makeshift community "pub." Extraordinarily, the fruiterer might have translocated from rural France, and after school he might teach the children some French as long as they did not get in the way in his busy shop. In 2000 the eucalyptus remain, with pendulous leaves that look as though they were dripping in the mid-summer sun, and the granite boulders as high as a house are still there, but the only remnant of the town is the store, boarded up and covered by lantana and morning glory. Farmers, the few still in the region, drive to the highway and then a further 30 kilometres or more to a roadhouse with fake aluminum countertops, gas, and fast food to pick up fresh milk and a newspaper. The population of rural areas in most of the "New Worlds" is one-third what it was early in the twentieth century; in 2000, the population of the whole of Saskatchewan, cities included, was less than it was in 1930.

The flat croplands and struggling towns of southwest Australia are punctuated by shimmering salt scalds and dead drainage channels (Fig. 62). Clearing has replaced the scrub with shallow-rooted crops that use less water; the rainfall has not changed, and so the water-table rises until it reaches the plant roots. What the pioneers did not know, and the scientists had not read about (but perhaps should have) was that this landscape once lay under the sea: thus, when the water rises it brings salt with it. In 1917, the local professor of agriculture, John Patterson, warned that part of this land would be too saline for agriculture; the crippled landscape of 2000 proves he was correct. Once the crops are dead or cannot be planted, the water budget shifts even more, so that nothing grows easily. Flat, salt-crusted lakes form and spread as though thrown about the landscape. For the whole of Australia, salinity was assessed in 2000 as affecting 5.6 million hectares of agricultural land, about 20 thousand kilometres of rural roads, and 68 towns. Modelling suggests that the numbers could rise, on a pessimistic estimate, to almost 14 million hectares, 67 thousand kilometres, and over 200 rural towns within fifty years.[12]

With respect to forests: "Only one-fifth of the world's original forests remain. Forest area has increased slightly since 1980 in industrial countries but has declined by almost 10 percent in developing countries."[13] By 2000, deliberate clearing of forests was continuing at substantial rates in Mexico, Brazil and Australia. In 1997 in Indonesia, drought and fires set by plantation workers destroyed somewhere between 150 and 300 thousand hectares of secondary tropical forests.

In Java, Indonesia, it is easy to see agriculture in action in the year 2000 as one crawls up a 30-degree hillside in a four-wheel drive too large for the narrow and poorly maintained road. Thirty years ago there were few people here; one might have seen only a few shepherds' huts among the trees, above the line where the soybean and rice planting stopped. But even then a traveller was never alone: in the seemingly remote landscape there would always be someone keeping the birds off the crops or walking across the ridges to the village. Nowadays people are everywhere, and it is a tribute to farming, and the fertility of the volcanic soil, that they are all well fed.

There is a price attached to the journey in time: forest clearing, by axe, tractor and, recently, fire has changed the landscape. The rice still grows at the bottom of the valley, where the water is more plentiful and can be held, at least for a few days, in the shallow handmade bunds (an embankment or dike) that have characterized Javanese agriculture since records began. Further up, there are soybeans, probably in the same places as before. Above them, now, are terraces filled with temperate crops such as peppers and cauliflowers. Once the higher land was not needed for food production and was too cold to reliably produce beans, corn, or rice; now, with all these people, it is cultivated continuously to provide cash from the sale of the temperate vegetables popular among affluent townspeople.

There are some encouraging data, too, in our millennial accounting (Fig. 74). In the United States, the total area of cropped agricultural land had not changed in twenty years, whereas the area of conservation tillage rose to 42 million hectares in 2002. In Australia, in the period from 1996 to 2001, sheep numbers declined by 30 per cent and rabbit numbers fell by as much as 90 per cent. Erosion remains massive, however, and is estimated to cost Australia $80 million each year in infrastructure repair, such as fenceline replacement, and $450 million in indirect costs such as water quality.[14] However, by the start of the new millennium erosion was probably, thanks to raised awareness, being reduced somewhat.

The key issues surrounding the maintenance of ecosystems and landscapes are recognized as complex. For example, in the State of the Environment Report for Australia for 2001,[15] the key issues were identified as: erosion, altered habitats, invasive species, soil salinity and acidity, nutrient and carbon cycling, and soil and land pollution. In 2000, the Australian government committed $1.4 billion toward regional and community work to reduce, or adapt to, salinity.

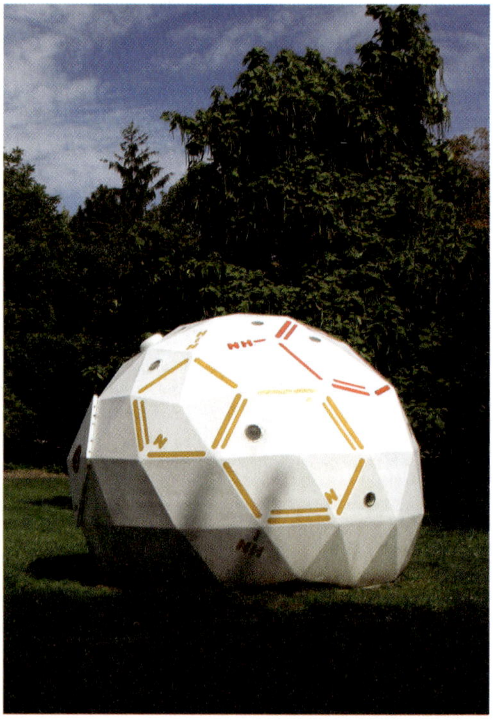

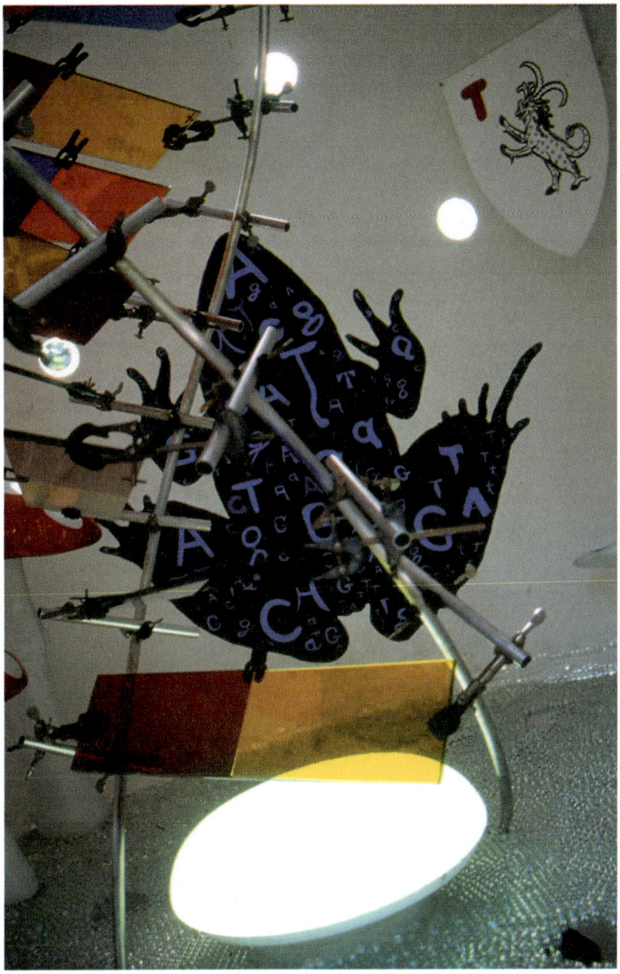

Opposite and above: Figure 74
FASTWÜRMS, *ex ovo omnia*, 2000 (mixed media). (Opposite) Exterior view. (Above) Interior view. The millennium was a time of accounting not only for scientists but also for artists like FASTWÜRMS, the collaboration of Canadian artists Dai Skuse and Kim Kozzi. In recognition of the year 2000, FASTWÜRMS created an imaginative sculpture in the form of a geodesic dome three metres in diameter. Situated in the Macdonald Stewart Art Centre's Sculpture Park at the University of Guelph, this work comments on the biogenetic engineering and biomedical research undertaken by University scientists. The name of this sculpture, which means "everything comes from the egg," is taken from an engraving on the title page of William Harvey's *Essays on the Generation of Animals* (1651). The exterior of the dome sculpture is imprinted with the letters A, C, G, and T, the symbols for adenine, cytosine, guanin, and thymine, the four building blocks that form the base pairs of DNA. Inside the dome the artists have portrayed medieval bestiaries on crests, and a strand of human DNA is represented by a double helix constructed of metal and stained glass. The interior of the sculpture also includes red-and-white fly agaric mushrooms. Giant images of a frog, a toad, and a falcon are imprinted with the A, C, G, and T symbols. The sculpture is humorous and intriguing, but also ominous in its suggestion of human–animal hybrids. Overall, the work presents a positive view of the potential for bioengineering to solve some of our future problems in medicine and food production.

Other issues, such as invasion by exotic species are recognized global problems, but they need to be addressed locally. For example, in the United States, the world's oldest national park listed its main issues as control or eradication of exotic species: non-native species of lake trout, and snails from New Zealand. Sadly, in 2000, despite rhetoric about biodiversity and the convergence of a number of countries toward signing a biodiversity convention (which would be done in 2002), there was little enthusiasm to spend money to protect biodiversity and habitats either directly or indirectly through, for example, eradication of the snails.

The start of the millennium has also been a time when governments have recognized that, as population increases, urbanization takes over alarming amounts of good-quality agricultural land. In the developed, affluent world this raises concern more in relation to landscapes and coping with atmospheric pollution; in the less-affluent world, as land is taken out of agriculture for houses, streets and shopping malls, it has an impact on food security. The numbers relating to loss of agricultural land give cause for thought. In Canada, twelve thousand square kilometers have been lost to urbanization in the three decades before 2001. Half of this land was classed as "dependable" – that is, in the better categories of land for agriculture. That area is equivalent to the size of Prince Edward Island. Although half of Canada's best (class 1) land is in Ontario, almost 20 per cent of this land, once used for farming, has been lost to urban sprawl.[16] The implications are well recognized: less land for food, lowered landscape aesthetics, and more runoff and heating associated with the tar of cities. Few governments have effective policies to deal with the issue of urban–rural landscape planning.

A review of the "state of play" with respect to agriculture and the valuing of natural resources and landscapes requires, finally, consideration of government corruption, government regulations relating to the environment, and public participation in agricultural and environmental development. This seems to be a mixed bag. On the one hand, governments in Africa, the poorest region, where agricultural improvement has been slowest and environmental degradation greatest, have in many cases become less committed to equity and the development of poor, often rural, communities. On the other hand, many governments are espousing the principles of sustainable development.

A most significant concerted action by governments that will affect agriculture and landscape management was the signing of the Kyoto Protocol to the United Nations Framework Convention on Climate Change, based on a script hammered out in Japan in 1997.[17] This called on governments to address four priorities: to enhance energy efficiency; to protect and enhance sinks and reservoirs of greenhouse gases; to promote sustainable forms of agriculture in the light of climate change; and to research, promote and increase the use of new and renewable forms of energy.

Eighty-four countries signed the Protocol that, as the words above indicate, is aspirational rather than concrete and binding. The "nitty-gritty" of what would actually be done and how it would be measured, at least for the first phase of the contract, took four years to negotiate until agreement was reached at a meeting at Marrakesh. By 2003, 109 countries had committed to action by ratifying or acceding to the Protocol. Significantly, while Britain, most of Europe, and Canada made commitments, the United States and Australia did not.

There is similar ambivalence in the international aid and development arena. International agencies such as the World Bank have put agriculture back on the agenda. Agency commitments to agricultural development, which were strong in the 1960s and early 1970s, loosened or ceased until about 2000. At this point, though, there was a rethinking. New policies were introduced that recognized that a healthy agricultural sector was important, and often a precursor, to urban-based community development. The ambivalence lies in the fact that government funding per capita or as a percentage of gross domestic income has declined for these agencies and for international research institutes, including the organizations that delivered the Green Revolution.

The World Resources Institute released a report in 2003 that indicated the widening disconnect between rhetoric and reality in participation: "Civil society groups are increasingly questioning the transparency and accountability of decisions negotiated by their governments internationally." This skepticism applies particularly to agricultural policies and "free" trade of food as poor, predominantly rural countries continue to get poorer.[18]

Finally, in considering the state of agriculture at the start of the twenty-first century, we should acknowledge society's ambivalence about the effectiveness and cost of government regulations. As with concerns that were expressed regarding top-down regulation for sustainability, there are always concerns, and publicly aired criticisms, of regulations. Increases in expenses, particularly unproductive ones such as accounting to a government, inevitably have an impact on the size of the workforce and on profitability. This argument was made and lost in Canada, or won in the United States, about the costs of implementing the Kyoto Protocol: Would costs to individual companies force them out of business, be passed on to the consumer, or cause technological innovation to create more competitive global positioning as well as improved agricultural practices for the environment?

And so, an accounting summary might read something like this: agricultural innovation and the rape of resources have reached a point where we have created a complex food system valued at $1.5 trillion that feeds almost 6.1 billion people, most likely exceeds the available natural resources of the globe, and definitely creates a momentum wherein the high-income countries are getting richer and the poor countries poorer. Within affluent countries, many farmers are among the poor and are getting poorer. The challenge for the future, which we will examine in our final chapter, is to find ways to address these issues, including by scaling back our use of non-renewable resources through connectivity and a new "regenerative agriculture."

7

CONNECTING THE FUTURE

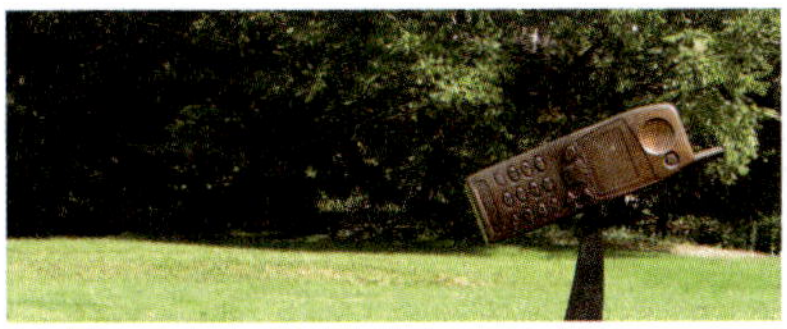

Where is agriculture heading? We suggest that its future lies in connectivity, and that connectivity may become the next paradigm, the next dominant way of thinking in many sectors of society, so that it lies alongside and eventually takes over from the paradigms of productivity and sustainability. Let us begin this discussion with three images, not of the future, but from today. They are from a college town in rural Ontario, but might as easily be in Toronto, New York, or London.

Image one: a glossy supermarket, isolated in a carpet of shiny cars and SUVs. Depending on the area of town, the aisles may be wide and the packages small, or the aisles narrow and crammed with cheaply packaged and super-sized standard fare. There is a large and slowly expanding area of produce labelled "organic" – expanding, because some customers are prepared to pay a small premium, perhaps ten or fifteen cents on the dollar, in the belief that they are supporting farming that is more environmentally friendly or because they think these foods are likely to contain fewer chemicals and are healthier for them. Elsewhere in the store, among the coffees and teas, there are options labelled "fair trade" for which customers pay a premium in the sometimes mistaken belief that they are supporting good labour practices and helping farming communities in far-off locales. Meanwhile, in the dairy displays, frustratingly mixed in with "normal" foods, are eggs and milk enhanced with omega-3 fatty acids, probiotic yoghurt containing an active bacterial culture, *Bifobacterium lactis*, and other such "functional foods" that back their claim of health benefits with medical evidence. The connectivities in the supermarket are direct and explicit (at least when the shopper reads the label): food is linked to environmental stewardship, community development, and personal health.

Image two: the Saturday morning farmers' market. In summer, stalls spill across the parking area with a dazzling array of vegetables and fruit, cheese, meat, maple syrup, bread, and home-baked apple pies. As fall comes, the stalls retreat and the range of vegetables narrows to assorted squashes. In winter the market is held indoors in a barn with foggy windows, offering scrawny vegetables and a skimpy range of pies, although the meat, cheese, and bread are still as good as ever. The seasonality of produce contrasts with the predictability of the supermarket, where it is summer year-round. Nonetheless, patrons of the farmers' market come for the sake of price, freshness, or the opportunity to talk with the farmers. These shoppers appreciate the "added value" of these stories and of the provenance of the food that they buy. This is a different kind of added value from the one gram of omega-3 polyunsaturated fatty acids in specialty milk, but is no less valuable. Some come, too, because they know the market supports a local farming community; after all, if the vendors are not supported through all the seasons, they may not be there when they are next wanted.

Image three: a farm about three kilometres north of the city. The creek is fenced off and bordered with a swath of grass and parkland-like trees; the house is in good repair; the fields are a tapestry of crops, grassland, and woodlot. A band of city-dwellers works on repairing a fence. To be truthful, this image is not real, but a composite of farms we know in upper New York State, England, France, and Sweden. The price of land is so high that no farmer could get a sensible return on investment from growing crop commodities. The key to retaining this land in agriculture is its connectivity with the urban communities, which value this piece of agriculture as an interesting "green space" that provides an aesthetic buffer against the city as well as providing clean water and air. City-dwellers visit and enjoy recreational benefits. The financial connectivity between the city and the farm is essential, otherwise the farmhouse would not be well-kept, or might not be there at all. Here, solutions vary with the local society: in Ontario, support comes from many small-pocketed individuals, coordinated through a church or land trust; in New York, it is from a large-scale, private, city-dwelling benefactor; while, in Europe, the connection is mediated by the state through taxes and payments for environmental services, based on the length of hedgerow or the area of grassland.

These three images illustrate that various societies have found ways of addressing "poor farmer, bad farmer" issues and supporting a "cultivated landscape." City people can and do come to terms with the complexity of our food system, our carelessness in using resources to an extent that exceeds their supply, and the international supply chains that seem to drive high-income countries to become richer while poor countries become poorer, just as urban societies in affluent countries become richer (or at least, consume more) while rural societies become smaller and poorer.

Where to? No one knows for sure; our future scenarios are no more or less valid than yours. Nonetheless, it is important to envisage what the future might be like in order to make a thoughtful, albeit small, impact on how it unfolds.

One desirable scenario could be described as "connected multiple futures." For developed, urbanized, affluent societies, important elements of this scenario will be the provision through agriculture of bioresources to replace the extractive industries that make one-time use of energy and minerals. Another key element for future connectivity will be the forging of common understandings between urban and rural people, and between societies in rich countries and those in poor ones. This effort will truly be the search for a "common sense." Once obtained, at least within regions, common sense will lead to new modes of landscape design. This new way of thinking, which emphasizes connectivity, is illustrated in a recent sculpture, *The Elevated Wetlands* by Noel Harding (Fig. 75).

But, before we venture into agriculture's new paradigm of connectivity, let us briefly review some of the fatal flaws in our existing paradigms. This exercise will oversimplify, but its purpose is to identify the gaps that need to be filled in with a new way of thinking about agriculture. And so let us consider, in turn, the paradigms of expansion, production, productivity, and sustainability.

The paradigm of expansion becomes untenable when we run out of space: there are relatively few unused arable spaces in the world that agriculture can exploit. Those that remain, such as the Brazilian rainforest, will be protected strongly by international opinion. This opinion is already loud in pointing out that forests are the world's lungs, are key in restraining greenhouse-gas induced climate change, and are precious for the biodiversity they maintain. Whether these views are equitable – why should the Brazilian economy pay for overconsumption and pollution in developed countries? – is not the issue. We might also foresee that, considered globally, the extent of land under agriculture will contract. The next fifty years may witness net reforestation for greenhouse-gas amelioration, to obtain carbon credits to meet international obligations, and to address soil and water degradation. Also, urban sprawl takes away the best agricultural land in many countries, covering it instead with ever-more-expansive houses and tarmac.

Productivity does not hold all the elements for the future, either. It has created a vicious trend, euphemistically called "declining terms of trade for agriculture," in which the relative price of food has fallen and continues to fall. Production of more and more food has been crucial to keep pace with the increase in population. However, the paradigm of productivity has two fatal flaws. First, by not fully costing or paying for inputs, it has caused agriculture to grow while food prices, in real terms, have declined. There are various examples of this, some of which were mentioned in chapter 4. Second, while food is cheap and surpluses are created for export in Western agriculture, the unit price of food at the farm gate is so low that farmers must receive other income from city-based populations to achieve a standard of living comparable to that of city folk. Cheap or underpriced exports in combination with import barriers in turn discourage food production in less developed countries, such that national investment is skewed and hunger and poverty continue unaddressed. Thus the paradigm of productivity, widely touted as the way for-

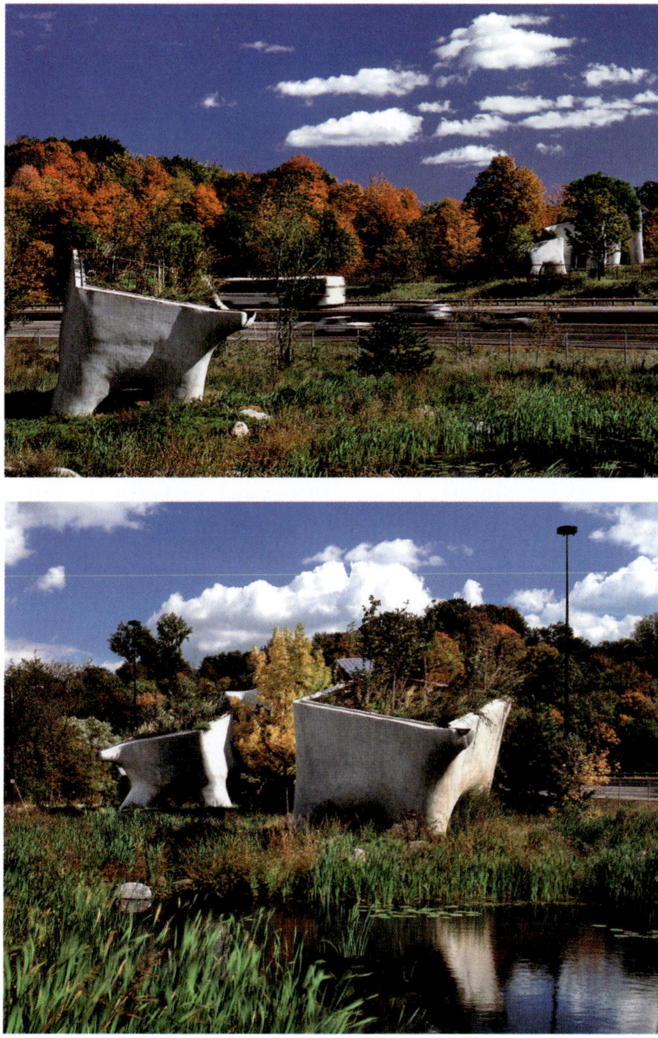

Figure 75
Noel Harding, *The Elevated Wetlands*, 1997 (garden material, concrete). Some late-twentieth-century artists have recognized and responded to environmental issues by creating utopian views, albeit with some irony. One of these artists is Noel Harding, based in Canada and the Netherlands. In 1997, he designed and constructed six demonstration gardens that were sustainable closed systems. Titled *The Elevated Wetlands,* the gardens are contained in massive three-metre-high concrete pods shaped like giant white teeth. The elevated wetlands are located on the side of the Don Valley Parkway, which streams thousands of cars a day from the suburbs into downtown Toronto. The gardens have an internal, self-watering system and provide sufficient nutrients for summer growth and successful winter dormancy. They are a plea to think about the current state of agriculture and the decisions we make daily that affect the planet, such as using chemicals for cosmetically enhanced lawns, supporting sustainable and organic farming through grocery store purchases, composting household waste, buying cars that use less gasoline, and supporting government policy initiative toward the new paradigm of connectivity.

ward for agriculture at least until the 1980s, has become known as the thinking that leads to "the race to the bottom."

The most recent agricultural paradigm – sustainability – is well-intentioned, but it will not be sufficient to carry us forward. After more than twenty years, it is a paradigm adhered to by citizenry, many governments (at least when it suits them), and international agencies and forums. But sustainability is only an outcome, an emergent property from a system once sustainability has been put into practice; in time, a particular agricultural system is revealed as sustainable, or not. History is littered with agricultural systems that, although encouraged by governments with good intentions, have proved to be economically unviable, to have caused depopulation of the countryside, and to have eroded the resource base. Good intentions and contemporary scientific advice do not necessarily deliver sustainable agricultural systems.

The definition of sustainability, the desired outcome, is also ambiguous. One-size-fits-all checklists are unlikely to be helpful except as a rough guide. For example, community-generated goals, which we have advocated elsewhere, may be entirely appropriate for making decisions about local development, but at a higher plane they are unlikely to engage and keep government support, nor perhaps address national goals. The maintenance of full employment, reducing odour pollution from agriculture, or reforesting the landscape are all laudable and achievable local goals, but a national program for sustainability must necessarily reach a view of the main goal, balance, or trade-off between them.

The recognition that, even with good intentions, one cannot always "get it right" leads some folk to the proposal that sustainability is an adequate paradigm as long as adaptive management can be used to achieve it. Adaptive management, sometimes referred to as a process and at other times as a paradigm, is seen by some as "a guiding principle for the interface between society and the biosphere."[1] It is based on incremental learning and decision-making toward agreed goals. However, it is applicable in a set of situations in which humans are causing environmental change, but in which there is time to reflect and to try various approaches: it is a process for adaptation, not for handling a crisis. Also, it works best, as does any process aimed at achieving sustainability, if the environment external to the situation remains relatively constant, or else changes predictably, while we search for sustainable agricultural systems.

Resource degradation and globalization (Fig. 76) are happening too fast to allow us the luxury of thinking that sustainability, implemented through adaptive management, will be good enough, or fast enough. In the present era, countries do unpredictable things, such as supporting global terrorism or declaring war on the pretext of

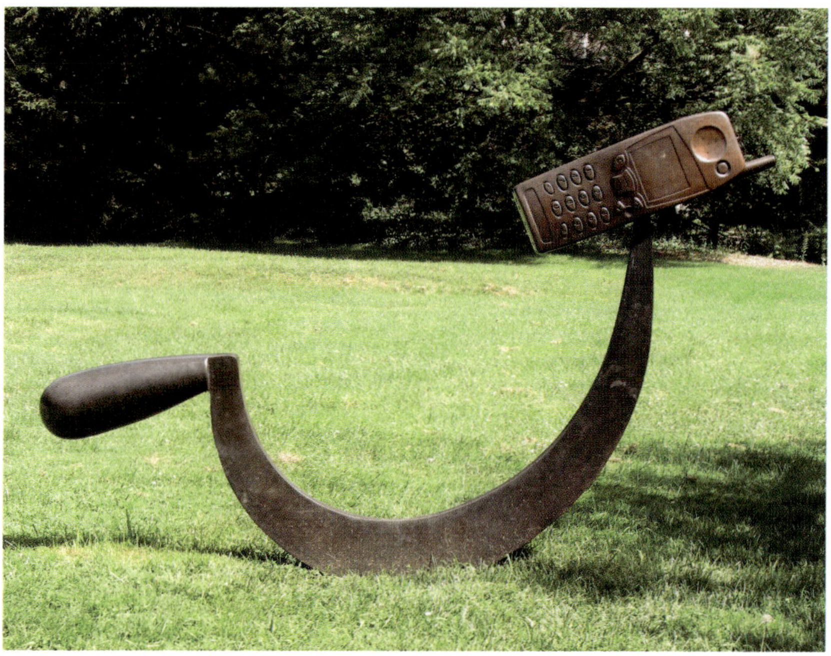

Figure 76
Gu Xiong, *The Sickle and the Cell Phone,* 2002 (bronze). Canadian artist Gu Xiong comments on the dramatic changes in China from the Maoist period to today, when the country is at the forefront of globalization. Xiong has remade the communist hammer and sickle symbol, replacing the hammer with a giant model of a cell phone. During the Maoist period, Xiong spent his teenage years working as a farm labourer deep in the mountains of Sichuan. Returning to the same region twenty-five years later, in 2001, he saw former peasants, now wealthy businessmen, brandishing several cell phones as they made international deals. He saw the demise of small centuries-old family enterprises, such as foundries, that could no longer survive. He also witnessed the displacement of masses of farmers to cities like Chongqing with its thirty-two million inhabitants. At another level, though, the juxtaposition of the sickle and cell phone represent the tensions, and the connectivity, within agriculture today.

unseen threats. Relatively few large companies shift resources quickly from one country to another. What is "sustainable" can not easily be judged if we rely on the global economy to create the rewards for local sustainability, especially when we cannot predict the actions of other national economies and companies.

Globalization is the growing integration of the world economy, made possible by information technologies that allow for rapid international financial transactions and enable production systems to become flexible and relocate where there is financial advantage. As a result, small national economies are more vulnerable to financial changes that occur elsewhere in the world. Trade and financial liberalization, such as the North American Free Trade Agreement (NAFTA), prohibits the regulation of capital through such things as former national laws that required capital investment to remain in a country for a minimum time. However, major capital flows are nowadays occurring in stock markets, where short-term return, not agricultural or environmental development, is the investor's objective. Speculative international flow of money, such as that which triggered the Asian crisis of 1998–99 and the Mexican crisis in 1994, has caused some economists to posit: "Capital market liberalization fails to distribute benefits and costs of economic integration for people in most need. Liberalisation is anti-planning ... [and] crowds out socially useful investment in favor of short-term returns."[2] As Joseph Stiglitz, who won the Nobel Prize in 2002 for his insights into markets with "asymmetric information," makes clear,[3] globalization may be a good as well as an inevitable thing, but its implementation, driven by ideologues from developed countries, is leading to more problems than solutions at this time (Fig. 77).

The hierarchy within sustainability – from field to farm to country to global environment – also conceals two problems. First, there is alignment. Social, economic, biophysical, and environmental issues and goals are rarely aligned in space, and only sometimes in time. Second, there is the paradox of empowerment: to initiate action at one level, say a catchment, routinely requires leadership or

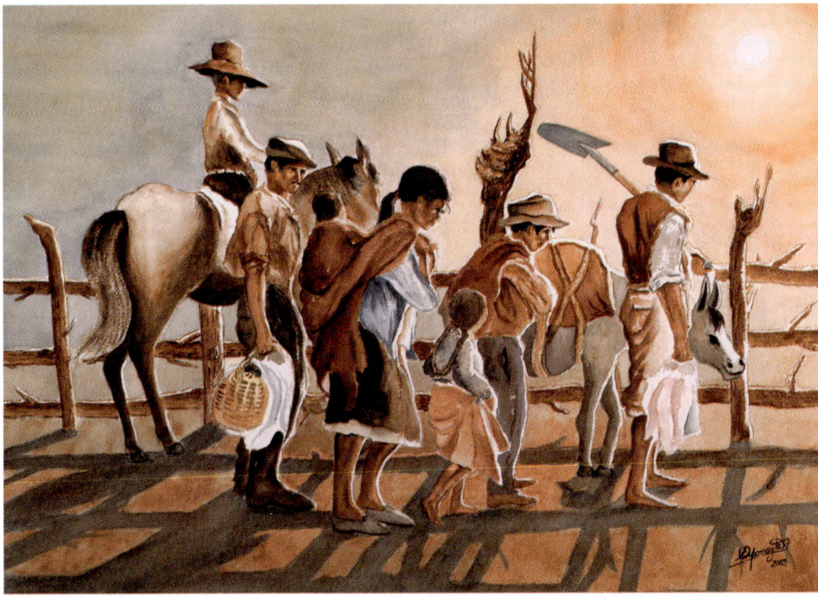

Figure 77
Ernesto Apomayta, *A Hard Day's Toil*, 2004 (natural inks on cotton paper). Peruvian artist Ernesto Apomayta calls for the modern adaptation of the best policies of his Inca ancestors, who built self-watering stone terraces, fertilized with guano. These terraces remain fertile today after 1,500 years of use.

facilitation and resources from the next highest level, such as the regional government. It is a variation on the problem of the commons: it needed supranational leadership to get an accord on climate change and, even now, if most countries do not abide by their commitments, it might be irrational for one or a few countries to make their contributions.

Food-system chains or "value chains" are trends that need to be accommodated in any view of the future. Once upon a time (in the nineteenth century before urbanization, and for the majority of the world's population outside big cities today), smallholders produced food and bartered or sold their surpluses in local markets. The next trend was toward centralized markets. Here, as in the Netherlands' central flower markets today, individuals or groups or cooperatives brought their produce to a floor and met potential purchasers. Each lot of produce was sold under separate negotiation. These have become increasingly sophisticated with neither the produce nor the purchaser having to be actually present as digital photographs are circulated and bids are taken from anywhere in the world; the produce is then shipped directly from the farm to the remote purchaser.

However, as the statistics from centralized flower and vegetable markets indicate, centralization and free-for-all selling and buying is giving way to integrated chains, in which wholesalers or retail chains make the purchase as part of longer-term contracts or understandings, either at agreed prices or on daily negotiated prices, depending on international markets. Farmers, or wholesale aggregators, bring their product to distribution warehouses or cross-docking houses. The adjective cross-docking reflects their activity: forty or more unloading bays accept shipments delivered in standard-sized trucks on prespecified pallets. The products are then placed in categories in shelves in stores covering 100 acres; within these, mixed loads are collected by forklift according to orders preprogrammed into their computers; the orders, once electronically verified, are shrink-wrapped in bundles on pallets to suit the retail store, then passed to other trucks that load and deliver, perhaps three times per day, in response to just-in-time ordering by the retail supermarket. The chains, and the supermarket experience itself, have a major influence on the type of product: consistency, quality, and reliability of supply are crucial.

These integrated chains source their products from anywhere in the world: canned goods from Spain will be on the same shrink-wrapped pallet as flowers from Columbia, bananas from Ecuador, oranges from California, apples from Chile, and wine from Italy and Australia. It is very likely that, within ten years, six global supermarket companies will dominate the sourcing, distribution, and retailing of the vast bulk of food for urban Europe, North America, and Asia, although it is not yet clear whether rising costs for energy will reduce both the global spread of sourcing – the geographical span of the "value chains" – and the reliance on just-in-time supply.

As one struggles to explain the flaws in our current paradigms of agriculture and to accommodate the rise of food-chain multinationals, one word keeps echoing: connectivity. We have not yet connected agriculture's consumption of resources with the value of those resources; we have not connected production with food consumption loci; we have not yet connected the need to reduce our ecological footprint – through greater use of renewable materials – with ramped-up agriculture for bioresource, rather than food, consumption. We are not adequately reconnecting the rural and urban spheres, whether to maintain the landscape in an arid Australian countryside or the rural culture that depends on high population density in sophisticated France, or to encourage self-development of viable rural economies in Africa.

Consider the hierarchy or nesting of issues within agriculture today and tomorrow. These are, from the global to the local: climate change; environmental services; the shift of developed countries to economies based on bioresources; the generation of good rural livelihoods in developing economies; globally-distributed value chains; and landscape design. These are interconnected. They might all be said to address sustainability, but that means different things at each level, and different things to different people. The single-minded advocacy of one term to embrace everything is likely to increase the noise of debate and the divergence of opinion rather than to provide a focus that can be embraced at all levels. We can, however, adopt a way of thinking that *connects* all the levels. The paradigm of connectivity accepts that the issues are nested and interrelated; it is a way of thinking about, and solving, problems (Fig. 78). New skills and new techniques will become apparent only when we think about how to achieve connectivity instead of focusing on a lack of sustainability or a lack of productivity.

Connectivity brings attention to transdisciplinarity. It addresses problems and opportunities of scaling: connecting between levels, within hierarchies and between short- and long-term outcomes. It also draws attention to the need to strengthen sector linkages: the connectedness between various economic systems, such as low-income and high-income countries and spatially separated sectors, such as urban and rural areas within one country. Somerville and Rapport describe the need for:

> transdisciplinarity in many problem-solving situations, particularly those relating to large issues on the sociopolitical agenda. These include the failure to cope with world poverty and growing inequities between rich and poor (within and between nations); the failure to achieve sustainable environments; the failure to provide all citizens with some minimally adequate standard of health-care, even in some of the world's richest nations ... While we still need a unidisciplinary focus in developing new knowledge, this needs to be counterbalanced by an equally vigorous effort to reintegrate knowledge, which requires more than simply evolving parallel streams of knowledge.[4]

The bringing-together of people from different disciplines and with different value systems (Figs. 79, 80) is essential to create connectivity and address the list of issues identified earlier. It is also essential to recognize that in addressing these issues – in designing the future – some skills that traditionally have not been valued highly will emerge as important, while others will become less significant. Futuring – the visualization of various scenarios for the future – will continue to develop. Back-casting, the taking of a future position and chaining backward through a series of plausible steps to the present, will need to become a more publicized, understood, and exacting field of scholarship.

The personal attributes of good management are likely to become more valued and sought after. Personal skills dominate a list of eleven things that have been identified as key factors associated with success, or with added value, in food distribution chains.[5] These attributes seem to explain much of why one chain – from consumer to primary producer – adds value and is successful while another withers or fails. They are: (1) awareness of relationships and needs among "players" in the chain, and particularly of the customers' needs; (2) trust, information flow and preparedness to undertake coordinated action; (3) business integration, usually stimulated by someone who is the chain leader; (4) efficiency; (5) customer focus; (6) transparency; (7) inclusion of rewards to drive success; (8) leadership; (9) planning; (10) attention to building and nurturing relationships, which might well be considered part of being a good leader; and (11) optimization of value chains, which depends, as in

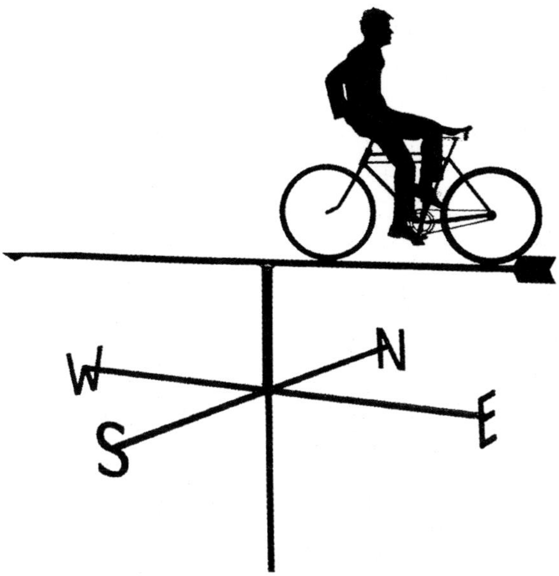

Figure 78
Rodney Graham, *Weather Vane*, 2002 (black enamelled stainless steel). Rodney Graham deconstructs a weather vane to create a symbol of our postmodern world, where there are no simple solutions for creating human connectivity with our landscape.

any business or aggregate, on ongoing strategic analysis, "keeping an eye on the numbers" to optimize performance.

Transdisciplinarity, participative processes and goal-setting, adaptive management, futuring, soft-systems leadership skills: it is a long and incomplete list of lots of fuzzy activities. We live in a complex world where we need not only reductionist science and economic skills but also the less clear-cut methodologies and solutions of complex systems. We have highly complex models to describe anything from groundwater movement to plant growth to global climate, and tools such as global positioning to bring our information to a coded spot on the land.

This leads to a paradox that needs to be remembered when we chart a way forward. Any of these complex processes or tools can benefit from refinement, and the scientists who spend years building computer models of crop growth, for example, naturally strive for that refinement, making the elements of their research and expertise more complex. The paradox is that, although refinement improves the ingredients, it does not necessarily make for a better meal: indeed, it is less, not more, likely to gain the attention of politicians and community leaders and to lead to action.

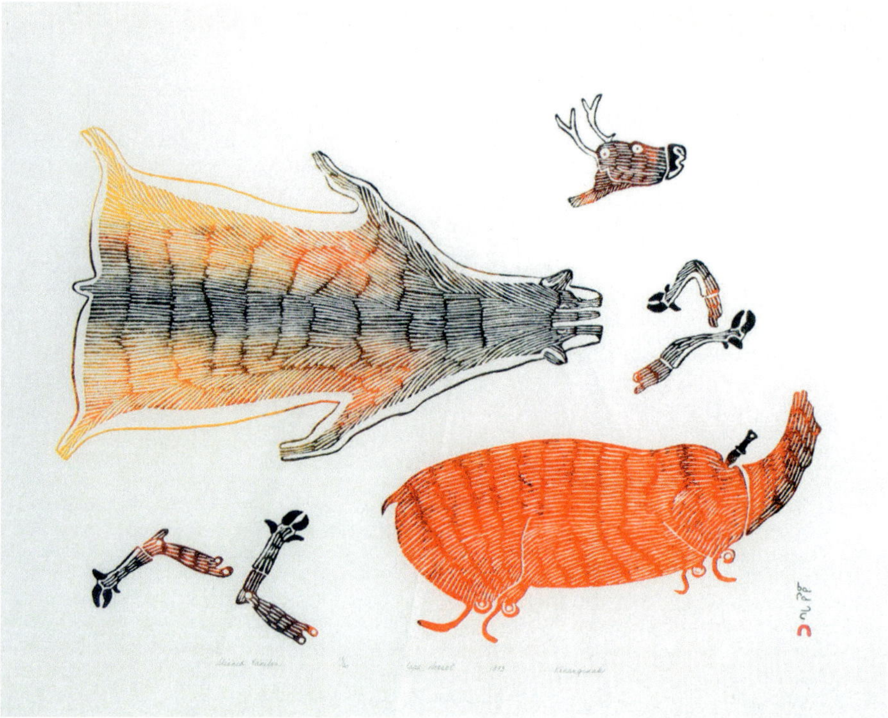

Figure 79
Kananginak Pootoogook, *Skinned Caribou*, 1973 (stonecut and stencil, 13/50). Although supermarkets have arrived in Northern communities, the traditional practices of hunting and fishing are still for many a preferred and more practical way of life. Pootoogook's stonecut celebrates a successful hunt and the nobility of the animal, and demonstrates that every part of the animal is used. A significant aspect of traditional hunting cultures is a view of nature as being in a continuous, cyclical union with humans. Traditionally, when a hunter shoots a caribou, a small act of propitiation is performed in gratitude to the animal and to quiet its spirit. Conservation is an important value in traditional hunting practice: taking only what is needed and can be used, even though today this may mean filling a freezer with cuts of meat. In the past, the Inuit faced periodic times of starvation as the caribou herds were depleted by disease and by adverse weather conditions.

Opposite: Figure 80
Emily Kam Kngwarray, *Emu Story,* 1989 (synthetic polymer paint on canvas). The connectedness of time and place is a major theme in the work of Aborginal artist Emily Kam Kngwarray, who did not begin to create art works until she was in her eighties. Her works focused on the desert environment that she called home and on the native animal species that co-existed on those lands. By layering dots and lines in a seemingly abstract work, she depicted the intricate paths of the emu herds, which are used as a food source by Australian Aborigines. For the majority of her 88 years she lived as her ancestors did, passing on hunting and gathering techniques to younger generations and the knowledge of where to find food in an environment that would seem desolate to other eyes. Her paintings encompass a world view which is shaped by the fertility of the land and by how it changes with the seasons.

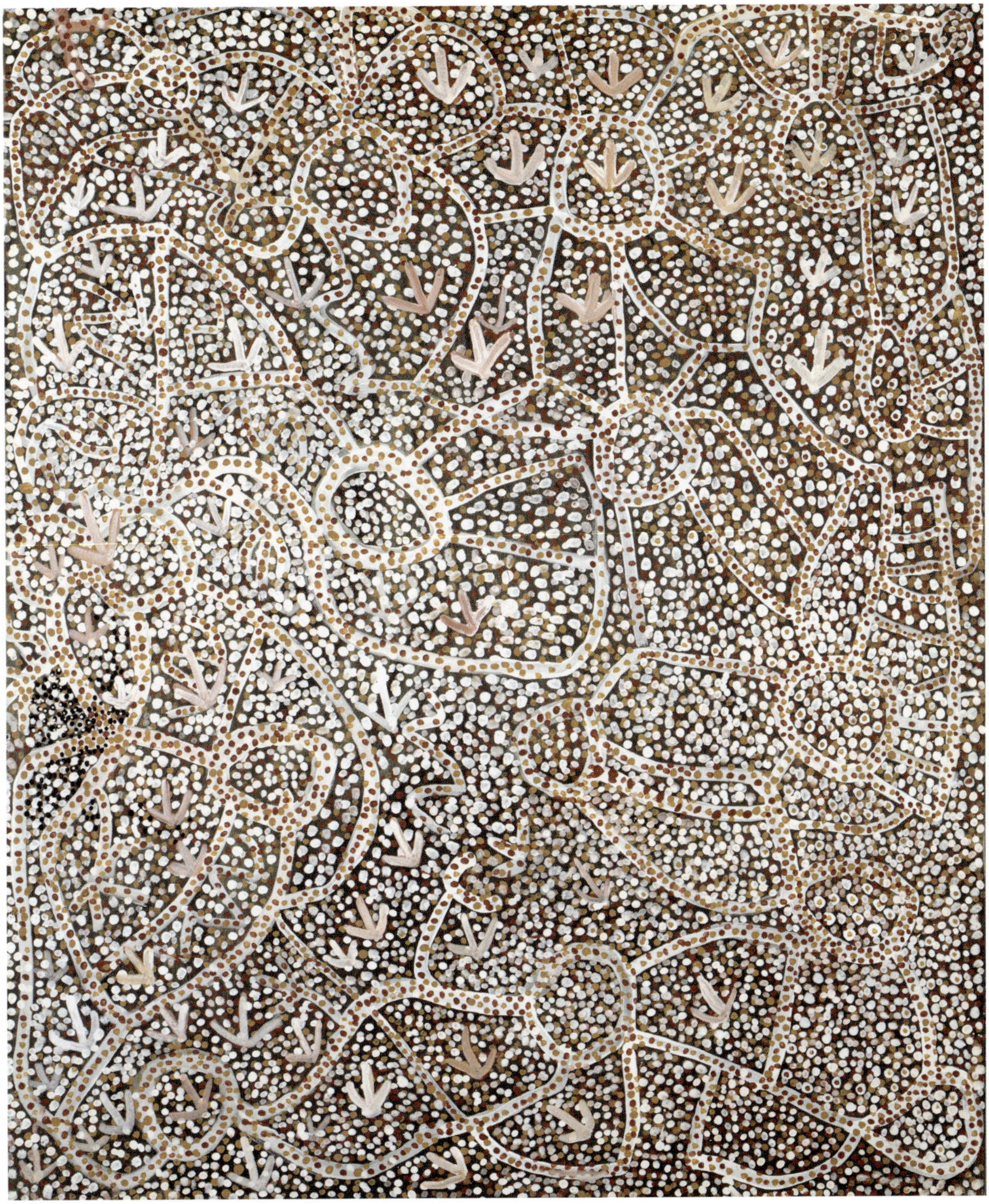

Political action on the issue of land degradation in Australia illustrates the pragmatic need for simplification. Scientists hold conferences on the many aspects of degradation that have been attributed to agriculture: soil compaction, loss of organic matter, acidification, and salinization. However, politicians look for issues that, if not one-dimensional, can at least be explained to the public and couched in terms such that they can be solved, or at least improved, in a short time. Simple issues, clear outcomes, defined program outputs and short delivery are the watchwords. Thus, it was David Dent and his colleagues who showed Australia's assistant agriculture minister of the day a ten-minute PowerPoint presentation to demonstrate that underground watercourses carrying salinity could be mapped, and that this knowledge could direct fencing and the targeted planting of trees, effectively achieving regional landscape design. This presentation triggered a $1.4 billion intervention to address soil salinity, starting in 2002. The engagement, and action, depended on a complex issue being presented simply. Could the money have been better spent in a program that addressed the complexities of not only salinity but also soil acidification and compaction? We will never know: the complex opportunity was never an option.

Climate change is the overarching issue that encompasses natural resource use and sustainability. Agriculture, food processing and distribution systems contribute to global warming and energy use, and hence play direct roles in reducing the generation of greenhouse gases, as well as having positive roles to play in creating "sinks" for trapping carbon dioxide.

However, as farm organizations are quick to point out, agriculture is often a relatively small contributor to the generation of greenhouse gases. In Canada, agriculture contributes less than 10 per cent to national greenhouse gas emissions, and land-use change and forestry have a favourable impact by trapping carbon dioxide. In Australia, by comparison, agriculture figures prominently in greenhouse accounting: 20 per cent of greenhouse emissions are attributed to agriculture, and land-use change and forestry contribute another 7 per cent, largely because of the continued clearing and harvesting of forests. To these numbers, because of the way countries have collectively agreed to carry out the accounting, we need to add a proportion of the emissions from the energy, manufacturing, and transport sectors as they move our food from field to fork.

The extent to which agriculture will contribute to reducing greenhouse gases will depend on technological innovation. For example, in Australia almost 70 per cent of agriculturally produced greenhouse gases are due to ruminants – mostly cattle and sheep – fermenting their food in their rumen and releasing methane. A low-cost vaccine might stop the production of methane and make a major contribution to reducing the national emissions.

In most cases, reducing greenhouse gas emissions makes good social and economic sense. As an example, elimination of land clearing in Australia would reduce emissions and shift that country to Canada's situation, where forests have a positive impact on greenhouse gas emissions. In the agricultural sector, too, emissions from livestock manure are the second-largest source of greenhouse gas emissions after rumen fermentation: manure accounts for about 13 per cent of Canadian agricultural emissions. Reducing these emissions, or trapping the gases for conversion into energy, would provide an immediate social benefit – reduced odour and less unsightly manure – and long-term energy savings. Ironically, despite the negative effects of industrial livestock production listed in chapter 3, it is the large-scale intensive livestock sector that is most able to adopt technologies that will address climate change. Their scale means they could have the capacity to implement low-cost vaccinations or to devise a system to trap methane and contribute it to a city's transportation or electricity grids, rather than allowing it to escape, as it must in rangeland ranching.

Extensive agriculture, and agriculture that is a component of broader economic activity (as is always the case), makes for more difficult innovation. The extent to which agriculture can reduce its greenhouse emissions depends on its size and on flow-on activities in local economies: innovation, either voluntary or government-led, depends on economic activity and the ease with which the innovation can be embedded within a system. For example, in Iowa, a midwest Corn Belt state that has always been identified as agriculturally based, agriculture in 2003 accounted directly for only 10 per cent of all jobs and 10 per cent of the state's gross product. However, every job within agriculture supports seventh-tenths of a job in the rest of the Iowa economy. But even when we track the jobs and value that it creates in other sectors, agriculture accounts for only 24 per cent of the state's total industrial output.

A global commitment to reduce greenhouse gas emissions will have the greatest impact on agriculture by triggering changes in livestock management and by encouraging the planting and management of tree-lots (Fig. 81), permanent grasslands, and the build-up of soil carbon – all of which will change the landscape. However, we need to be realistic: there are many small family farms and small rural towns in the "Iowas" of the world that will find it difficult economically to reduce greenhouse gas emissions from agriculture. For example, being paid to grow tree-lots to capture carbon may sound attractive, but in good agricultural regions such as Ontario the areas are small and the likely value of carbon credits too low to persuade farmers to shift from higher-value crop production. In areas of lower productivity, such as the North American prairies and the Australian wheat belt, shifting to low-input forestry may be economical.

Although it is important to take steps to slow climate change, a second, relatively unrelated issue is to adapt to changes that are already in trial. Virtually all models indicate that the world, regardless of any action we take, will continue to get warmer, particularly through a raising of minimum temperatures; that it will become drier in parts; and that the weather will become more variable, with increasingly frequent and severe storms and droughts. This area of agricultural science and soon, management, is called climate adaptation. It is already clear that the impact of climate change on agriculture will be regional, and that some areas will have, for example, higher minimum temperatures, drier seasons, and lower crop yields, while others, perhaps nearby, will experience different impacts from climate change. Thus, predictions of what the future of agriculture will be like – the mix of crops and overall regional productivity – depend largely on scale: that is, on whether calculations are made on an average regional scale or by summing up the effects, both negative and positive, of adaptations at a smaller scale. The calculated average regional effects seem more benign than what is likely to occur if we model adaptation at local levels and add these. For example,

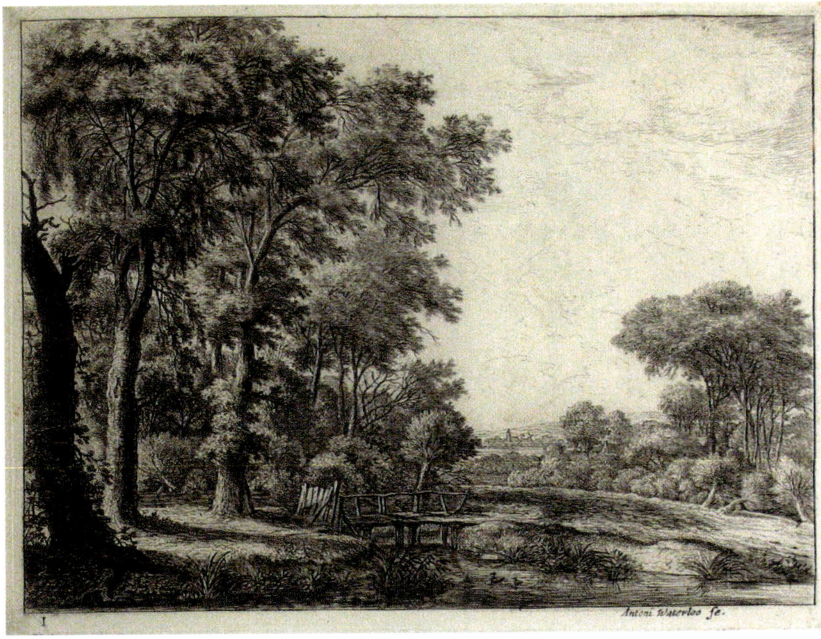

Figure 81
Antoni Waterloo, *The Entrance to a Forest with a Small Wooden Bridge*, circa 1660 (etching). Waterloo depicts woodlots integrated with fields on a grand estate. The seventeenth-century sensibility was to express a sympathetic union between humans and nature in landscape design. This approach promoted biodiversity and would be an appropriate philosophical model to guide thinking today. "Nature was alive but not a threat, as it tended to be for settlers in America. [Seventeenth-century] prints are an important visual reminder of how far we have moved away from life in balance with nature. Who would have thought, looking at this not-so-distant world, that in a little over three centuries there would be overpopulation, wide-spread environmental degradation, and global warming threatening the very continuation of life? There is no going back, but we need reminding of how far we have come."[6]

in the southeastern United States, climate change will have a negative impact: corn, soybean, and sorghum cropping, and agriculture overall, are likely to diminish while cotton cropping might increase.[7] Of course, the actual impact of climate change will also depend on how quickly farmers make management changes, such as changing the time of planting or the species of crop they grow.

Parallel with the need for farming to adapt to climate change is the need to shift to lower-input or more sustainable agriculture and to forestry or native vegetation reserves in developed, large-footprint economies. This adaptation will be accelerated if it is motivated by payments for environmental services as well as products such as greenhouse carbon credits. Here, connectivity is important: the goals that motivate landscape change need to be aligned among the global, national, regional, and local spheres.

One practical way in which agriculture can take the lead in landscape design, which aligns global, national, regional, and local needs, is to consciously design our future cropping systems. Semi-closed or "regenerative" systems can be designed, and optimized, to maximize local recycling and take account of the true costs (in terms of energy as well as the financial costs to society) of farm inputs such as fertilizers and pesticides. Regenerative systems can also be designed to minimize, indeed eliminate, unwanted outflows or pollution from farming. The need for, and characteristics of, regenerative systems would become a key to implementing connectivity.[8]

To achieve connectivity there are crucial issues on which we need broad definitional agreement. Public goods, environmental services, and land rights are three of these that are crucial. Public goods are benefits which extend beyond the individual or group that undertakes the activity that creates the benefit. In our context, public goods that result from farmers' actions are conservation or rehabilitation benefits such as reduced salinity, improved water quality, enhanced biodiversity, and aesthetically pleasing landscapes.

A centrepiece of economic and policy thinking of the last 20 years has been the centrality, and dominance, of the corporation. Ronald Coase (b. 1910), who won the Nobel Prize for economics in 1991, advocated the centrality of the company in playing the key role of lowering transaction costs – the "noise" or waste of doing business. Until the pendulum began to swing again, this perspective led to the valuing of property rights and to the view that externalities or public goods should be minimized and handled, where possible, within the private sector.

In agriculture, property rights are a key issue in future landscape design. Farmers generally take a stance that their right to be free of interference from government should exceed that normally enjoyed by urban property-holders. As previously rural areas become admixtures of urban and rural lands, and as urban sprawl relies heavily on rural land managers for clean water and green space, this view has become untenable. It is nonetheless important that rural landowners have some certainty about their ability to act or, conversely, about the likelihood of government intervention. The reason for this is that farms, unlike urban homes, are business units: without reasonable long-term certainty, farmers will prefer to make short-term investments, and in so doing reduce society's capacity to address the issue of sustainability.

A recent parliamentary enquiry in Australia[9] into "public good conservation" identified several principles. These began with landowner rights in respect of land use: that is, not necessarily the right to unfettered action, but some certainty about long-term opportunities that would colour farmer investment. The enquiry also agreed that an individual such as a farmer has a duty of care to manage land in a sustainable manner. This duty may be mandated, but this committee, at least, advocated setting a "baseline" as to what should be expected at the local level. A third principle, which is gaining widespread (if grudging) support, is that repairing damage is a shared responsibility. This means that in high-income, highly urbanized societies, remediation will involve shifting money from cities to rural areas.

What are appropriate funding mechanisms to address agriculturally based environmental issues? In all developed economies, there seems to be broad community consensus that these mechanisms should involve urban–rural connections: city money is needed to contribute to addressing environmental public goods, or externalities, even if they must be addressed by farmers in the context of landowner rights.

Four mechanisms can be used to achieve this. One is government-supervised transfer payments. Since the ratification of the Maastricht Treaty of 1992, the European Union had provided billions of dollars to farmers for two purposes: sustainability and multifunctionality. The latter refers to things that are created or preserved other than goods for trade: landscapes, culture, recreation, and the preservation of the rural way of life. The United States government, through its Farm Bill, provides billions to its farmers to conserve – not farm – highly erodible and environmentally sensitive areas, and targets funding toward the creation of conservation reserves and the adoption of conservation practices within farmed land. In Australia and Canada there is intermittent public discussion about an "environmental levy" on all taxpayers, such that cities would contribute the most tax money to address environmental externalities in rural areas.

Agriculturally generated "public goods" or externalities do not necessarily require direct tax support, though. Two other local ways are private-company and community provision. Government policy-setting to encourage land trusts or payment for specific services, such as clean water from particular catchments, employ market, corporate-oriented economics to achieve public benefits. The growth of The Wildlife Trust and the Woodland Trust in the United Kingdom, the Audubon Society in the United States, and the Australian Conservation Foundation (among other organizations) in Australia all suggest global applications for the problem. Communities of common concern, whether registered or voluntary groups, address conservation issues. By 1995, it was estimated that 37 per cent of Australian farmers were members of Landcare groups. A parallel may be drawn between these groups and other types of associations – say, a health club, a privatized urban water authority, or a neighborhood grocery-purchasing collective, through which many individuals pay for amenities that they would not be able to provide or afford on their own. Whether such associations make a profit depends on their goals, efficiency, and governmental or legislative context.

In general, as the group that has responsibility for "ownership" and implementation becomes less connected – more remote – from the entity that sets the goals and provides the money, so the wastage is likely to increase and the impact to diminish. Thus, while the Landcare movement is laudable on many grounds, such as creating commitment and changing the value-systems of rural people, its effectiveness and efficiency were questionable when it implemented a billion-dollar government program that was loosely defined and was left largely to groups for self-monitoring.

A fourth way to compensate farmers for providing environmental services and goods – cleaner water, for example – is through the marketplace. Here, multinational corporations increasingly seek an advantage over their competitors by advertising that their products are produced sustainably. "Fair trade" tea and coffee are examples: consumers pay a small premium for goods produced according to standards that give a fairer return to the producer, do not use child labour, and encourage sustainable, environmentally friendly production practices. The Mennonite Central Committee is credited as having initiated the fair-trade movement in 1946, but the movement became "mainstream" in Europe in the 1960s with respect to handicrafts. About 2000 fair-trade products entered the food market, beginning with tea and coffee. An umbrella organization established in 1997, Fairtrade Labelling Organizations International (FLO), inspects and certifies producers in more than fifty countries. In 2004, the FLO was divided into two different organizations: FLO International e.V., which sets fair-trade standards and provides producer business support, and FLO-CERT GmbH, which inspects and certifies producers. This is an example of a third-party accreditation system that, although responding first to terrible worker conditions and child labour, also addresses the need to protect the environment in poorer countries. In 2005, it had an impact on over one million farm families and workers with sales of more than 660 million Euros, trickling more money, perhaps 10 cents a kilogram, back to farmers and helping to connect agricultural production with better care for the environment.

A second example of paying for environmental goods and services though the price of food is not mediated by a third party such as FLO but through action, and advertising, by the producer itself. This tends to be restricted to large multinational corporations because of their advantages of scale and their large advertising budgets, but increasingly we see smaller, boutique producers "bundling" their food products in connection with claims of care for the environment. Bananas provide an unlikely example of this bundling, or connectivity. Global production of bananas, a fresh fruit produced nowadays on a gigantic scale on tropical plantations, is dominated by two vertically integrated companies, Dole and Chiquita. Like Dole, Chiquita's predecessor, the United Fruit Company, created a low-cost global advantage through clear-felling of forests, aggregation by purchase of small holdings, and creation of large plantations based on a narrow genetic stock – within cultivars such as "Williams" and "Gros Michel" – and intensive methods that required lots of cultivation and chemicals. The lack of genetic diversity in the plant stock together with the production systems meant that Panama disease became rampant and, as in the early age of expansion (chapter 2), production declined over time despite increasing inputs. Fields were abandoned, workers were left without jobs, and villages that had grown up near the packing plants had no reason for being. In 1991, a non-government organization called the Rainbow Alliance proposed to Chiquita a "Better Banana" and a more environmentally friendly system of production. As Craig Canine wrote in 2006:

> By the end of 1995 every Chiquita banana farm in Costa Rica had been certified by the rainforest Alliances' growing team of independent auditors. The improvements in Chiquita's operations in Costa Rica – cleaner farms, less labour strife, and greater operating discipline – made the bosses at company headquarters in Cincinnati take notice. Word came down to roll out the Better Banana program to all Chiquita farms in Central America and by the end of 2000 every one of them had earned certification.[10]

Banana plantations still operate on a scale that is difficult to imagine: they roll across horizons, and the packing sheds are still like huge aircraft hangars, ill-lit to make it easier to keep the fruit cool and slow its ripening. But, whether we are buying a food that is prominently labelled "organic" or "fair-trade" or a brand name like Chiquita, which has internalized some environmentally conscious production practices, we are contributing to the explicit connectivity of agriculture and the environment.

Another way of ensuring payment for environmental services and sustainable farming practices is through the food distribution chain. Here, multinationals increasingly seek a market advantage over their competitors by advertising that their product is produced sustainably. In-company accreditation and monitoring is becoming commonplace. Although this may ensure that, for example, Europeans eat Chilean apples only if they have been produced without soil erosion, it is unlikely to address externalities: the lack of a price premium attached to sustainable production suggests that chain-driven payment for environmental services will need to be leveraged by governments, rather than using price premiums to solve the problem.

When we consider mechanisms for funding the environmental services that agriculture provides, these options lead to consideration of how policy can be best set to leverage or encourage private investment (in context of a business, seeking return on investment) and benefaction. The answers will vary from region to region. However, a checklist of impediments to private investment bears repeating:[11]

- Lack of clear plans. Diffuse, complex issues need simple objectives analogous to, for example, a project proposal for mining investment.
- Low rates of return. This may change if we begin valuing the public costs, such as the costs of moving a town affected by salinity.
- High risk. Agriculture is seasonal, and prices are seldom controlled by the farmer unless he or she is creating a unique or premium-quality product.
- Illiquidity. Investors are tied to a natural resource project that is not portable, and they may be unable to enter or exit.
- Small unit size. Generally, land trusts require effort to aggregate holdings to achieve scale and deliver outcomes.
- Externalities and "free riders." As the benefits will not be captured by a few, rewards must recognize the value in terms of public good.
- Few existing institutions. Very few listed companies specialize in this area, and lending institutions have little experience in assessing natural resource and rehabilitation projects.
- Limited government support. Paradoxically, in those countries that have high levels of support or subsidy, this is not helpful to investment. A large proportion of the money farmers receive from subsidies and "safety nets" is ploughed back into land prices, which make it more difficult to enter agriculture and tougher to obtain a reasonable return on invested capital once one is in the farming business.

- Complexity.
- Poor information. Although mainstream investments, such as mining, are supported by detailed and independent analyses, this has yet to become common in agriculturally based resource management.

As an aside, both greenhouse accounting and environmental services raise the need to measure, or develop indices of, vegetation. In the case of greenhouse gas emissions, this is required under the Kyoto Protocol and will also be stimulated by markets in carbon credits. Environmental services or multifunctionality, as practised by both the United States and the European Community, have found this measurement and monitoring too much trouble, although pollution – from agricultural nutrients or by other means – has demonstrated that this laxity is not philosophical, merely pragmatic. At any rate, these assessments for environmental or conservation purposes are mandated in some legislation, as in the case of Australia's National Forest Inventory, which requires each state government to undertake a detailed assessment of the extent, biomass and floristics of its forested lands after five years. Such assessments will become commonplace once people invest serious money in environmental services through taxes, targeted environmental levies, or personal investment in companies. There is an enormous number of indices and procedures, spanning the spectrum from the simple-and-useless to the impossibly complex and scientifically rigorous. As an illustration, the health and public value of a forest may be given a score from zero to 100 on the basis of criteria of area, fragmentation, mortality (whether the canopy is thinned by bugs or fire, for example), and erosion. These may easily be estimated from satellite images and knowledgeable, but cursory, inspection on the ground. This index, applied to four 10,000-hectare tracts of Brazilian rainforest, gave values of 98.5, 88, and 68 from imagery in the years 1972, 1986, and 1992, respectively.[12]

This index-rating of the Brazilian rainforest illustrates one of the issues we have yet to confront: that the valuing of a landscape depends on the perspective or level from which it is observed. Is the forest of international significance as a carbon sink, in which case it might be logical for the international community to contribute money for its preservation? Is the forest rare and endangered, or simply a good example of a landscape that is already well represented? These sorts of questions lead to the notion of priority bioregions and representative vegetation types, which has emerged since about 1990. It includes the idea that the world's undisturbed vegetation – that is, pre-agriculture, if anyone can reconstruct a past that far distant – may be classified, and specific regions or countries have a responsibility to maintain examples of their native bioregions. National reviews of biodiversity may be superimposed on regional priorities to identify and preserve a representative system of biogeographical or ecological regions.

The selective preservation of representative pieces of biogeographical regions superimposed on local landscape design is a national or international obligation. Richard Thackway[13] has enumerated steps in this process: (1) to identify gaps in the current suite of protected areas; (2) to identify sites that may fill the gaps; (3) to select potential reserve areas; (4) to assess their feasibility and negotiate their inclusion into protected areas that would be excluded from agriculture; and (5) to establish government-gazetted reserves and plans for their management. It is encouraging that the carbon accounting that will be undertaken for greenhouse emission monitoring will use these "interim biogeographical regions," at least in Canada and Australia, where the classifications are best developed.

Future government policy responses to climate change, together with adequate valuation and payment for environmental services, will create a connected framework within which a country's agriculture will develop. It is likely that we will see a rapid divergence in agricultural activities according to country. Broadly, the trends will be toward bioresource-based economies in countries with large ecological footprints – that is, the high-income lands – and a preoccupation, at least through our lifetimes, with generating good (or, to be more honest, adequate) rural livelihoods in low-income economies.

Within high-income, energy-consumptive economies, there will also be divergence. This variety will create "multiple futures." Holistic and organic agriculture continues to grow and is likely to reach

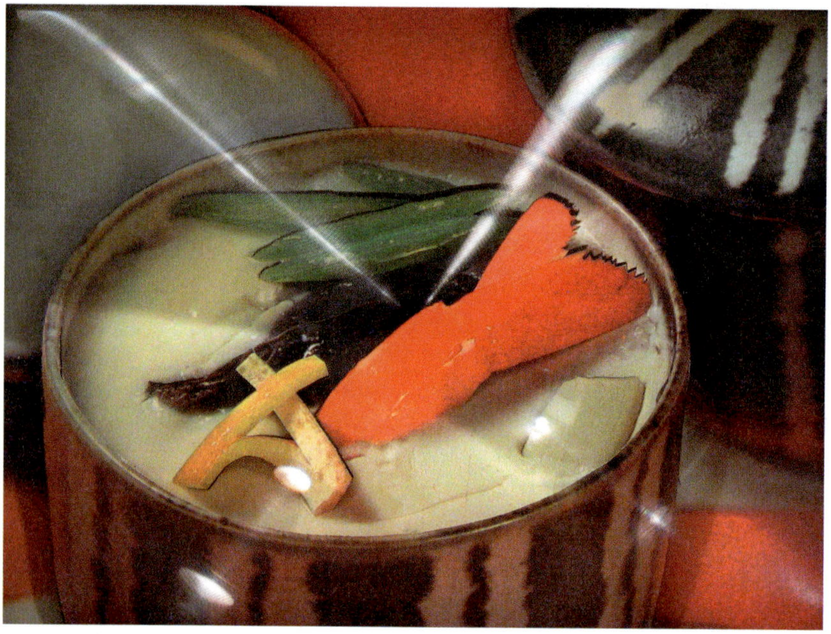

Figure 82
David Hoffos, *Steamed Egg Custard with Vegetables*, 2005 (shadow box with laser print and magnifying lens). Hoffos comments on the ersatz nature of processed food, whether from the take-out window or row on row of supermarket freezers. His supposed still-life photograph of an Asian delicacy is in fact extracted from a Japanese recipe card book originally published in 1983. Hoffos commodifies the food as object, and distances it from its intended instructional purpose: to put food on the table.

one-third of the sector by 2010 in the European Union and perhaps in countries such as Canada and Australia; traditional agriculture will develop with and without genetically modified crops, with segregated supply-chains for each. Intensive industrial agriculture for low-cost, high-volume production seems likely to continue to serve, at a minimum, the North American fast-food market (Fig. 82).

In addition to the futures already mentioned, crops are being bred and management technology developed to produce naturally occurring pharmaceuticals, food additives with special nutritive properties (e.g., anti-oxidants), car fenders, as well as products such as cardboard, food wrap, and housing materials that are already on the market. Ethnobotanical research may also lead to new – or rediscovered – applications for cultivated plants (Fig. 83).

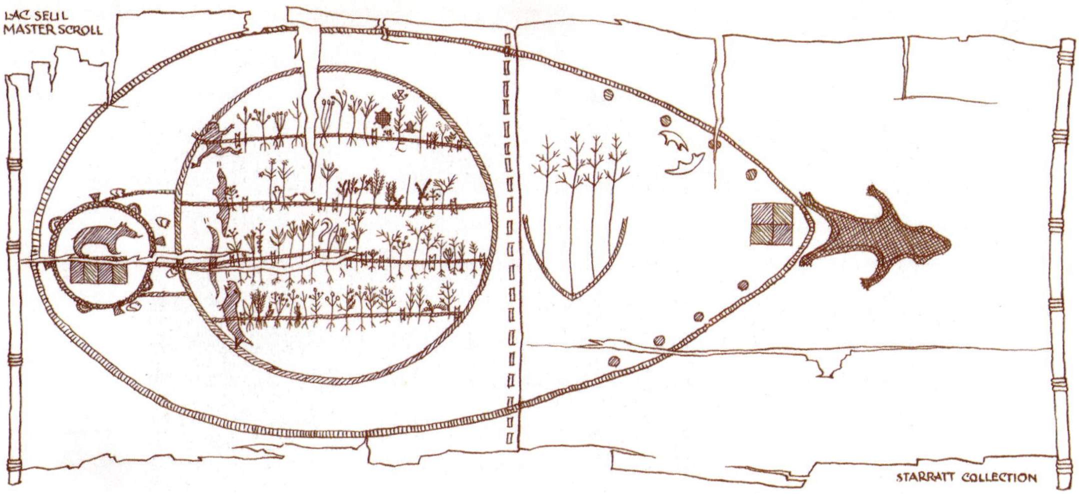

Figure 83
Southern Ojibway, Artist unknown, *Lac Seul Master Scroll with Bear emphasis and herbal references*, not dated (natural pigment on parchment). The sacred birch-bark scrolls of the southern Ojibway people contain teachings about native plants used in ritual and for medicine. More scientific investigation of native people's botanical applications is needed to save vital, centuries-old information before it passes away with the present generation of elders. This knowledge could lead to new agricultural crops or the introduction of conservation regulations for forest plants found to have medicinal value.

The bio-resource-based economy, dependent on agriculture, will require realignment of land use and connectivity between agricultural land use and industry. The exacting quality needs of the processing makes it likely that a few growers, under contract, in the best agricultural regions will supply product to new nodes of processing plants. Like today's grape or tomato growers, they will be close to the processors; equally, processors will create clusters as some ingredients go into biotech incubators and others, which once would have been waste, are made into newspaper, clothes, or wallboards.

However, world population growth will not stop, and the challenge of feeding the 800 million who are underfed continues. The Green Revolution showed it could be done with high-yielding crop varieties and higher levels of non-traditional inputs. Today's assessment is that, although feeding the people is an essential first step, it is not enough. Hence, there is a search for ways – and there will be many – to improve food production and at the same time improve the self-esteem and the livelihoods of the rural poor, particularly in Africa. The challenge in low-income countries is exacerbated by the tragedy of the HIV/AIDS epidemic, which has particularly affected women. The population of sub-Saharan Africa in 2020 will be 38 per cent lower than estimated earlier, because of the disease.

At a meeting of the Global Forum for Agricultural Research in Dakar in May 2003, there was a consensus that "good livelihoods" were the goal, but agreement on how this was to be achieved, except in the most general sense, was vague. Some themes did emerge, however, which might be grouped into three categories.

First, there is continuing emphasis on participation and empowerment. Local farm families have to adapt, not adopt, technologies for their goals. Collectively as villages, they can set medium-term local goals, such as buying a threshing machine or a brick-maker, to set the village on the path out of subsistence farming. There are many examples of community development through self-empowerment. Kristin Helmore and Naresh Singh detailed two in *Sustainable Livelihoods*:[14] villagers in Gaviotas, Colombia, stimulated by an expatriate, developed appropriately low-tech technologies for farming, coupled with arts and crafts, for wealth-generation; in Gandali, Malawi, development came through the marketing of briquettes made from crop residues and grasses that, an attractive product at a third of the price of firewood, brought cash into the community.

The notion of collectively developing sustainable livelihoods need not be confined to low-income communities such as these, however. In the United Kingdom, farmers' organizations and churches have joined forces to encourage farmers to come together and brand their farming activities in a way that provides them with a set of values and guidelines for production, and that also might give them an identity and some power in the market. The LEAF (Linking Environment And Farming) movement in the United Kingdom began modestly in 1991. It has 3000 members and advocates "commonsense farming." Its mission statement illustrates the interconnectedness of issues associated with modern agriculture:[15] "LEAF is committed to a viable agriculture which is environmentally and socially acceptable and ensures the continuity of supply of wholesome, affordable food while conserving and enhancing the fabric and wildlife of the British countryside for future generations." By 2004, LEAF was supporting 45 demonstration farms throughout the United Kingdom. These were designed not only to attract farmers, but also to educate urban families, the consumers, and local law enforcement, who will increasingly determine the role of agriculture within urban societies. Their attraction is strong: "See how (empowered) farmers are balancing environmentally responsible farming with running a profitable business."

Empowerment also implies that farmers get adequate prices for food products and keep some of the profits that are generated within multinational chains. Western communities have a responsibility to address both. The price of food is distorted by artificially low-cost grains from North America and dumped food from Europe. This can be changed. Further, the sharing of benefits within global food distribution chains is conceptually easy: indeed, it is one of the tenets held by advocates of value. It is quite possible that affluent consumers, attracted to one retailer or another on the basis of their claims to ensure that their food is produced under socially and environmentally sustainable conditions, can actually ensure just this, through public exposure of low-income farmers' conditions and competitive shopping.

Another category of change that seems necessary for rural development is illustrated by the experiences in Gaviotas and Gandali. Rural communities need connectivity with urban wealth. The noble, subsistence farmer does not represent the future. Youth migrate from rural to urban areas, where they usually earn higher wages. Take the example of farmers who are migrating from the countryside to Chongqing, the largest city in the world. The farmers leave their small villages where they practice traditional organic farming and arrive in the city with little more than a stick and a rope. With this simple tool, they become the movers who transport masses of food and commodities on their backs each day. They live in city-provided dormitories and send money back to their families.

Money needs to be returned and reinvested in agriculture and near-farm processing. Infrastructure, together with mechanisms such as cooperatives or multinational retailers, need to be encouraged to move the agricultural product from the village, in top quality, to the market. This flow, from high-income urban markets to low-income farmers, is the opposite of what prevails nowadays: when low-income countries are repatriating up to 50 per cent of their gross domestic product to pay interest on overseas loans from high-income countries, it is not surprising that even rudimentary services such as health are declining and there is no money left over to leverage private investment in agriculture.

Rural development depends on good governance and access to science and technology, not ideology. Several African countries have the resources, technology, and labour to achieve sustainable, empowered development; others have abundant natural resources such as energy reserves that would, if the returns were invested wisely, create "lift-off" without resource degradation. At the Dakar meeting, in a glossy hotel conference hall holding four hundred participants, the President of Senegal exemplified the triumvirate of needs in his address to the meeting: local empowerment and adaptation; urban–rural links and the support of global financial systems; and good science and government.

Global climate change, continental resource valuing, payment for ecological services, and national and regional agricultural production are all connected. And, as discussed, they might easily be aligned. Our goal is to create and maintain viable social, economic, and environmental systems across regional landscapes. Thus, at the regional and local levels, it is landscape design that has the exciting and exacting challenge of delivering them all. It is a challenge that applies in both rural and urban spheres. Frederick Law Olmsted's plan for New York City's Central Park (1858) was an integrated design that successfully blends meadows, reservoir, active and passive recreational areas, buildings, and woodlots. Olmsted's design was imaginative and promoted biodiversity on a relatively small site (Fig. 84).

In the present day, the challenge intensifies: the hill we climb is always getting steeper. Population growth creates a need for intervention and constraint; otherwise it will, through spread, have an impact on protected areas. Populations are becoming ethnically and

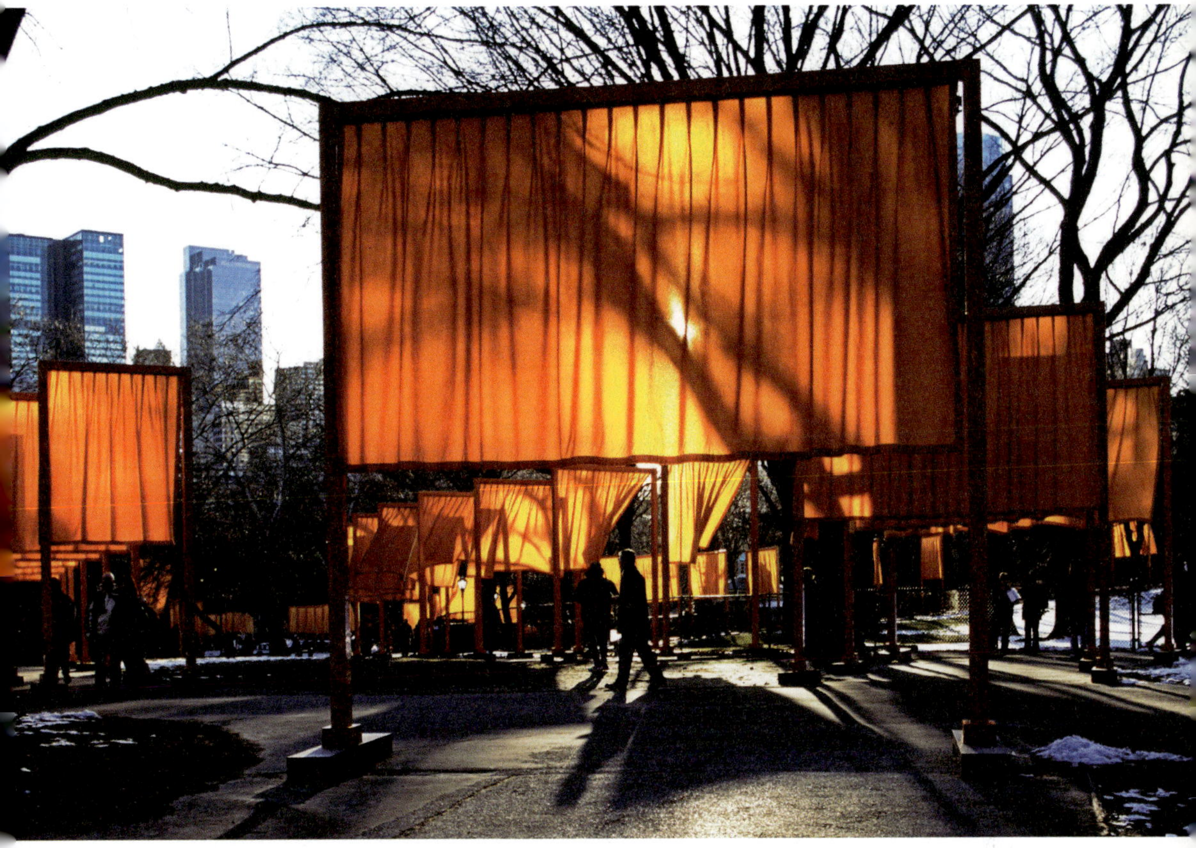

Figure 84
Christo and Jeanne-Claude, *The Gates, Central Park, New York City*, 1979–2005 (steel, vinyl tube, cast aluminum, nylon thread). This public artwork reminds us of the value Central Park holds for New Yorkers. After twenty-five years of lobbying by the artists, City Hall finally agreed to allow them to install 7,503 gates of flowing orange fabric throughout the park's walkways and trails. By imposing their art work on the garden created by Frederick Olmsted, the founder of the landscape architectural profession, the artists reminded New Yorkers of the incredible asset the Park is to their lives.

socially more diverse, with different value systems, needs and opportunities. Economies, and specifically food chains, are becoming global and are beyond the influence of local policies.

South Florida provides, in microcosm, an example of the increasing complexity of landscape design and the increasing need for it. South Florida is in the Sunshine Belt, and is a point for immigration, particularly from Latin America. It is literally running out of land. Redevelopment to increase population density is seen by government and the community as necessary to accommodate the next wave of new residents. It is essential that this redevelopment is planned and constrained if urbanization is not to cause further environmental damage to the Everglades. This is a protected area, the only subtropical reserve in North America, and was declared an International Biosphere Reserve in 1976 and a World Heritage Site in 1979. Plans for urban redevelopment to address these environmental needs must simultaneously take account of the complexity of the communities that it is proposed to "redevelop": issues such as cost of housing, culturally preferred urban landscapes, and attachment to place all need careful consideration.

Visiting the Everglades reserve, we were greeted by a group of six or ten Seminole elders, all male, surrounded by younger villagers. With a mix of pride and embarrassment, they showed us the community properties, the cut-up fields lined with stones and stakes, the two cows, and a good-looking, sleekly brushed calf standing quietly near the timber and iron shed. We edged to the open brown dirt space where we squatted and talked about what their ambitions were: where they wanted to improve things, if such improvement was possible, and where they were content. We asked quietly whether the women would be joining the conversation, and let it pass; it would be another two visits before the women joined in and discussed their own priorities, such as milk for the children and reading.

Participatory planning in rural development and in designing more satisfying landscapes is much the same whether it occurs in the highlands of Ethiopia, South Florida, or with a group of farmers in a mechanized, Western society. The local values and enthusiasms need to be identified and built upon by the local people; change may be assisted by expatriate technocrats, extension officers, or consultants, but if these outsiders drive change, history shows, it will collapse when they leave.

However, in this century, unlike in all previous ones, scientists and economists can bring powerful and flexible information support to the design process. Geographic information systems containing layers of data from several disciplines can be brought to the discussion. This can, in "real time" during a conversation, be manipulated to create visual representations of regionally relevant options for land use. As an example among many in a rapidly expanding field, future landscape scenarios have been created for the American Corn Belt so that farmers and policy-makers can consider, in three dimensions, what a now-familiar landscape would look like in 2025. What if, in pursuing a goal of increasing agricultural production, they favoured ranching and closer farm settlement to improve water quality and flows, or enhanced biodiversity while maintaining agriculture through perennial strip intercropping and agro-forestry?[16]

We are entering an new era of landscape design. Sometimes it is by local choice. More often it is forced upon local communities because there is "nowhere to go": they may have already run out of land, like the residents of Florida, or cannot produce enough food, like the Ethiopian highlanders. It is widely accepted that the process of design needs to support, not direct, those who will continue to live in the landscape and bring the design to reality; sometimes, however, broader national or regional goals will create a directive or compulsory planning framework for agriculture, as commonly occurs in urban development.

Effective design is assisted by three things: (1) participation and ownership, which is today taken as a "must" but is sometimes still overlooked by the leaders of the process; (2) technical support, such as spatial information systems that allow scenarios to be developed and made "visually real;" and (3) alignment between social and ecological spaces or regions. The last-mentioned has received relatively little attention, although it is widely acknowledged that national and regional government boundaries often bear no relation to natural aggregates of people: their ethnicity, religion, social activities, or even, locally, their shopping patterns. Some studies have described these "social networks" or spaces, and recommended radical realignment of boundaries, such that local government area boundaries would connect with social networks on the one hand and with natural agro-ecological boundaries, such as a watershed, on the other.[17] Whether this sort of permanent restructuring is necessary is moot; in any event, recognition of the importance of connectivity is emerging as another "must."

Western agriculture is entering a new paradigm, which we call connectivity. It will not replace but, rather, will add another dimension to the existing paradigms of productivity and sustainability. Connectivity will include, commonly, landscape design. The landscapes will be increasingly diverse and divergent: we are creating "multiple futures" that will include forests and grasslands managed for greenhouse accounting and the preservation of biodiversity through representative biogeographical regions, organic and low-input agriculture, and conventional techno-centric agriculture for bioresource-based manufacturing in which plant products are synthesized into food, wraps, car parts, and high-input industrial agriculture.

We are, through intentional design, recreating the landscape as a work of art or of purpose. In what the aboriginal people of Australia called the dreamtime of their ancestors, before the coming of Europeans and their agriculture, mythic animals such as the rainbow serpent meandered across the landscape, laying eggs and gouging out beds in which to sleep. The hills, rocks, valleys, and trees are what we see and respond to aesthetically, and they are also holders of myths. Figure 85 depicts a serpent, its body disguised by reed-like appendages so that, when it slept in low-lying areas, it would be mistaken for, or became, swamps covered by grasses and reeds. The emu crop on the rainbow serpent's chest is where it stores the water and hail that it spews out to create the wet season in Northern Aus-

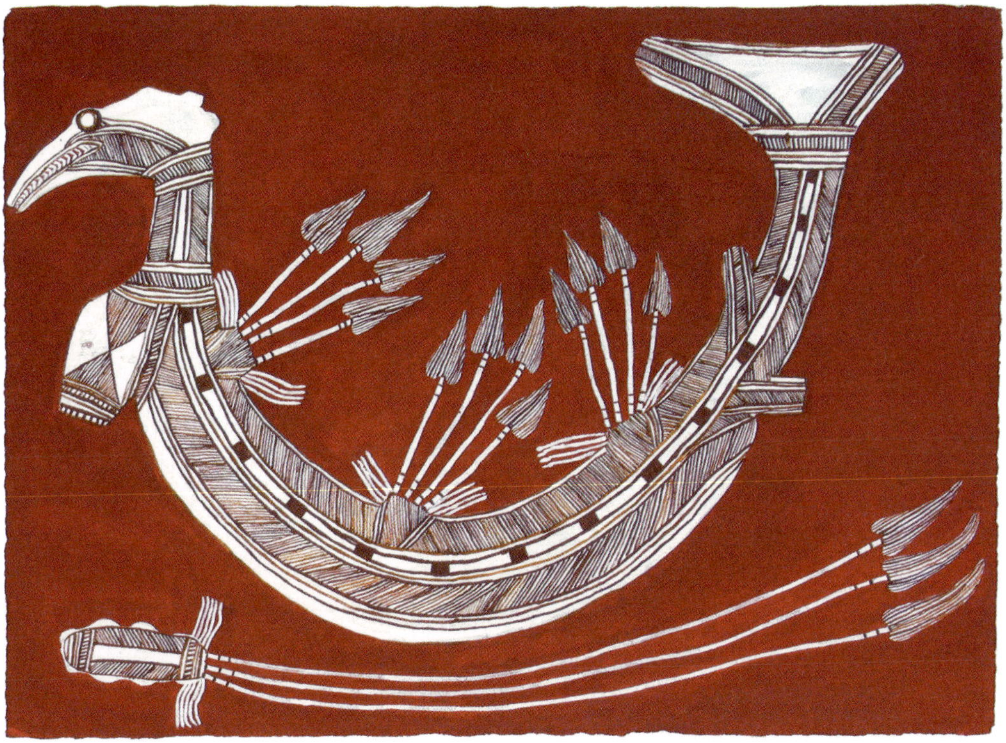

Figure 85
Bardayal (Lofty) Nadjamerrek, *Ngalyod, the Rainbow Serpent*, 1998 (natural ochres on paper).

tralia. Aboriginal storytelling is a way of communicating about humans' place in the natural order of the universe – essential knowledge we have chosen to ignore.

Aboriginal stories and art frequently address the seasons and the rhythms of growth, expressing a connectivity between people, landscapes and their spiritual world. Traditionally, Wandjina paintings (Fig. 86) were repainted annually to ensure rain or the increase of an animal species. The repainting act was a significant ritual linking humans and nature. Contemporary versions of the ceremony provide meaningful links for the artists to the animals that represent clan totems. The enduring relationship of humans and nature is also expressed, but in a more abstract way, in the drawing *Willow Bushes* by Inuit artist Irene Avaalaaqiaq (Fig. 87).

These works by aboriginal artists powerfully reinforce the argument that, after sustainability, society needs to take a "connectivity turn." It is powerfully ironic, however, that the very aboriginal people who, through their art work, cause us to reconsider our connections with the land and our food, did not generally practice any form of crop or livestock husbandry. It is to the hunter-gatherers and nomads that we turn for inspiration and metaphor as we seek futures for highly urbanized, agriculturally dependent civilization.

We have in this book journeyed from the highly-structured manorial past through two hundred years of expansion and increasingly cut-throat "productivity," a paradigm that is increasingly being referred to as the "race to the bottom," to arrive recently at a society that speaks in terms of "sustainability" but has not yet the tools, and perhaps not yet the will, to implement this new paradigm as a "way of being." Further, environmental issues – and the excessive "footprint" of resources that are being used by a few affluent societies – make it clear that we do not have the luxury to continue with business as usual. Our faith that the present economic system coupled with the paradigm of sustainability will maintain global civilization will not deliver us from agriculture's "dark corners." A way forward is to give agriculture the attention it needs – it is the basis for most civilized societies – and to think "connectively." We need urban-based societies to pay the true cost of food and create realistic prices, both to maintain countrysides and to stimulate food production in less developed continents. We should demand agreed methods of valuing the environment and intangibles such as quality of life, and follow this with electoral support for local government intervention to create environmental design. These are charges for change that fall predominantly on urban societies.

Changes in urban responsibilities must be coupled with demand from urban societies for change among farmers: for farmers to take personal and collective responsibility for moving from the image, which their own organizations often promote, of being "poor," and the image that urbanites often attribute to farmers, of being "bad,"

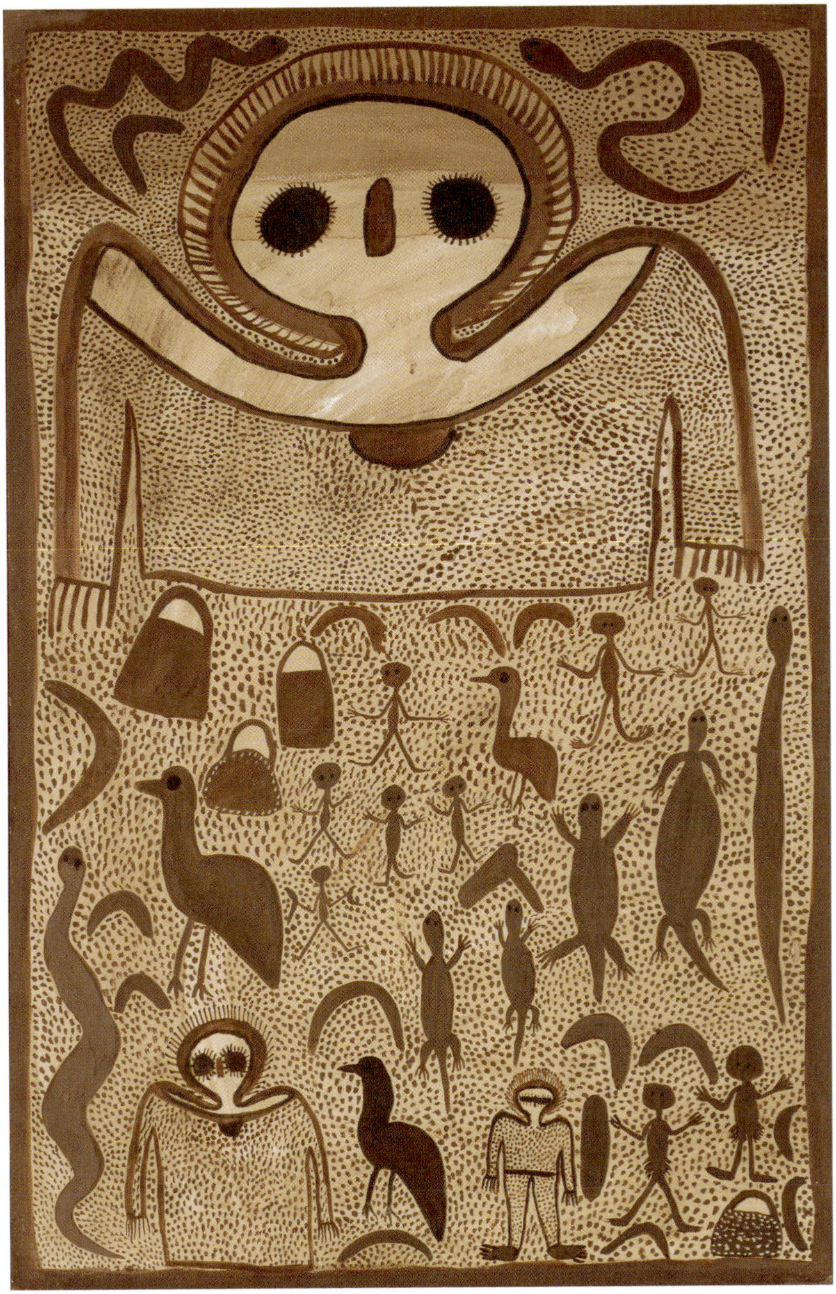

Figure 86
Lily Karadada, *Wandjina,* 1990 (earth pigments and natural binder on canvas).
The *Wandjina* is a powerful spirit figure that has eyes and a nose but no mouth.

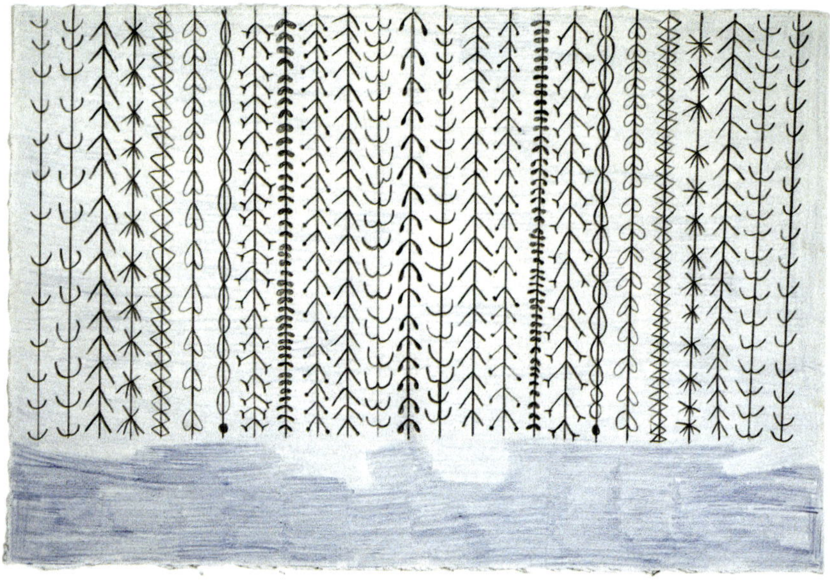

Figure 87
Irene Avaalaaqiaq, *Willow Bushes,* 1970s (graphite and coloured pencil on paper). This drawing depicts the endless cycle of the seasons and the plants that sustain the caribou the artist hunts. Avaalaaqiaq explained that the arrows pointing up indicate the plants in the springtime and the downward arrows show the plants at the end of their growing season. The colour blue indicates the water, and the colour white indicates the snow. This highly symbolic drawing by an artist who was born and spent her early life on the land, renews her intimate relationship with the tundra.

so that they become again "good." Our future needs transformational regenerative agriculture to minimize our industrial footprint and develop high-value bioresource-based products: corn for cosmetics, health cures, and car parts. We have come a long way; the knack will be to find agreed future paths that are both informed by the past and transformational so that they create a new, connected civilization:

> Time present and time past
> Are both perhaps present in time future.[18]

APPENDICES

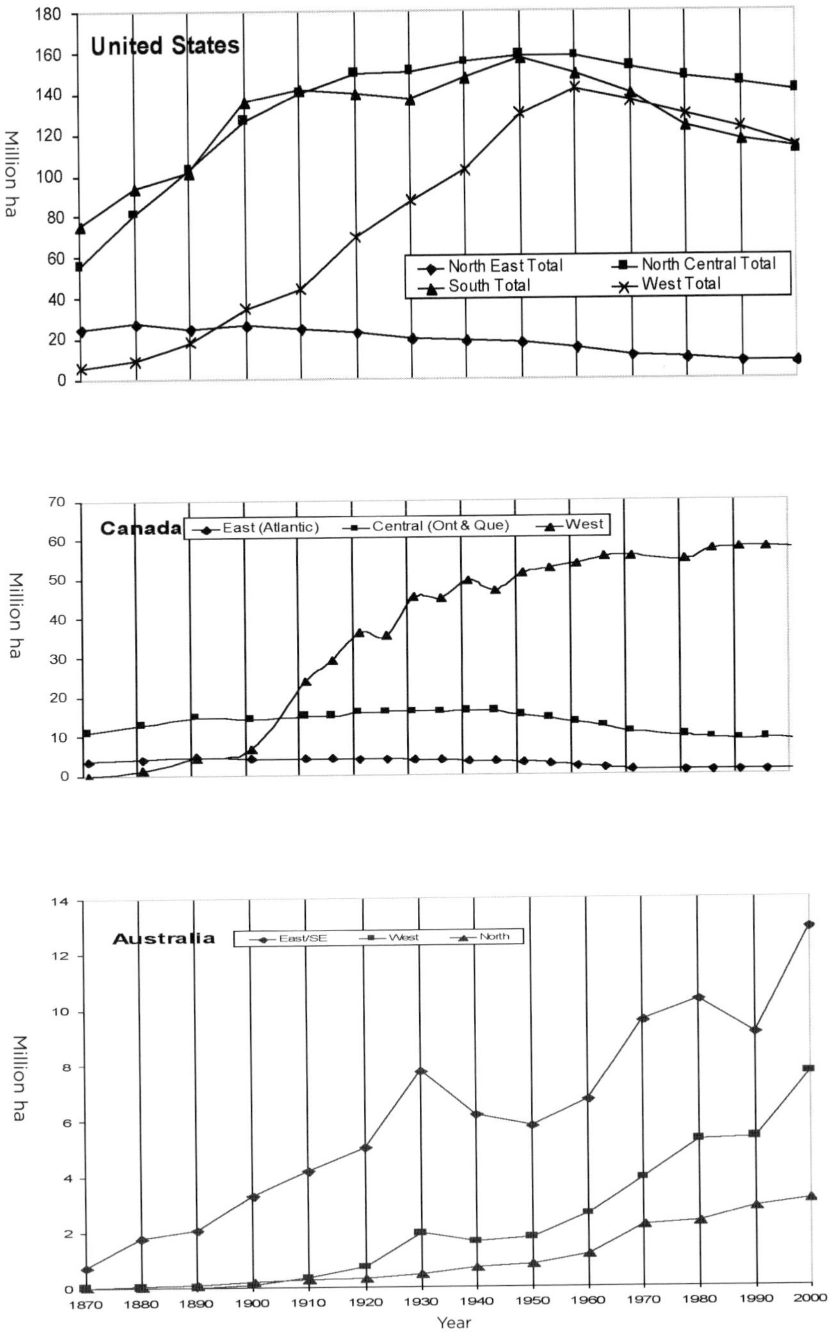

Appendix 1
Areas of farmland in Canada, the United States, and Australia, 1870–2000

UK Fertilizer Consumption and Total Wheat Production

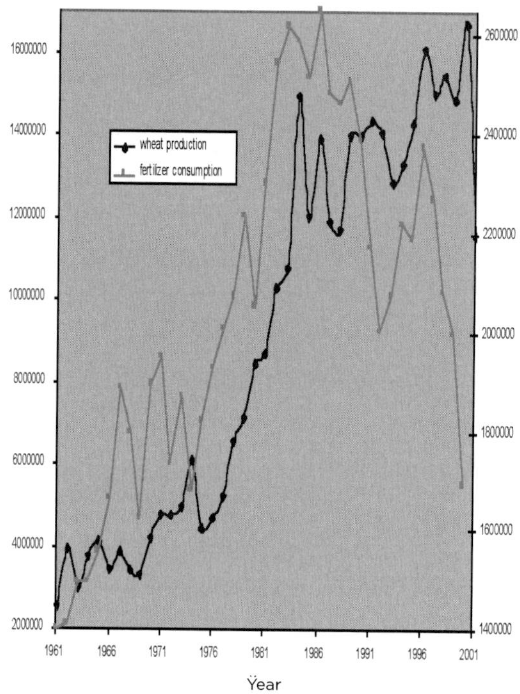

Appendix 2:
Fertilizer consumption and wheat production in the United Kingdom and the Netherlands, 1961–2001 (tonnes).
Source: FAOSTAT agricultural data.

Netherlands Wheat Production and Fertilizer Consumption

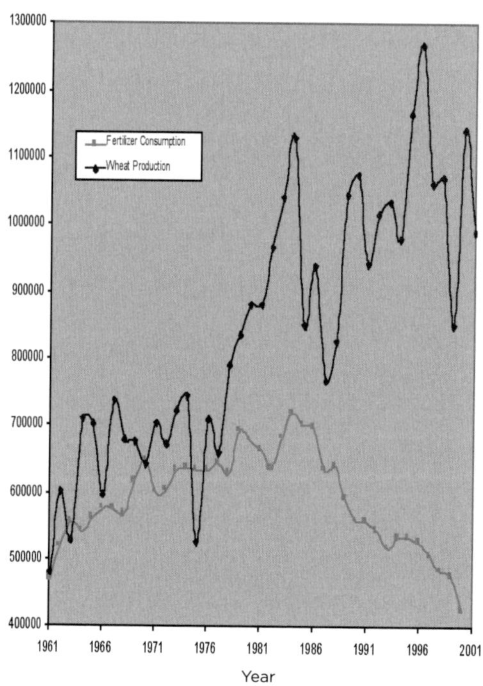

Table 1

*World livestock numbers**

Livestock	1961	2001
Chickens	3,891.4	15,301.6
Cattle	941.7	1,359.6
Sheep	994.3	1,036.9
Ducks	193.5	1,021.9
Pigs	406.2	924.8
Goats	348.7	735.2
Turkeys	130.7	251.0
Geese	36.6	238.4

*millions

Source: FAOSTAT agricultural data

Table 2

Aggregate of world production and area harvested of major crops

Crop	Production (tonnes, millions)		Area harvested (hectares, millions)	
	1961	2001	1961	2001
Bananas	21.2	67.9	2.0	4.4
Barley	72.4	144.1	54.5	56.2
Maize	205.0	614.5	105.5	139.1
Rice, paddy	215.6	597.8	115.5	151.2
Sorghum	40.9	59.5	46.0	44.2
Wheat	222.4	590.5	204.2	214.8

Source: FAOSTAT agricultural data

NOTES

CHAPTER ONE

1 Philip Clarke, Brian Jackman, and Derrick Merce, eds. *The Sunday Times Book of the Countryside Including One Thousand Days Out in Great Britain and Ireland* (Great Britain: Macdonald Futura Publishers, reprint 1980), 35, 248.
2 Cited in W. Rosener, *Peasants in the Middle Ages* (Urbana: University of Illinois Press, 1992), 212.
3 Rosener, *Peasants in the Middle Ages,* 217.
4 Lucius Junius Moderatus Columella, *On Agriculture and Trees,* trans. E.S. Forster and Edward H. Heffner (Cambridge, Massachusetts: Harvard University Press, 1954–55), 435.
5 Andrew Petzold, *Romanesque Art* (New York: Harry N. Abrams, 1995), 96–9.
6 A.K. Mehrotra, *Middle Earth* (New York: Oxford University Press, 1984), 40–1.
7 R.A. Butlin, "The Enclosure of Open Fields and Extinction of Common Rights in England, c. 1600–1750: A Review," in *Change in the Countryside: Essays on Rural England, 1500–1900*, ed. H.S.A. Fox and R.A. Butlin (London: Institute of British Geographers, 1979), 65–82.

8 J.I. Little, "Agricultural Improvement and Highland Clearance: The Isle of Arran, 1828–29," in *Scottish Economic and Social History*, 19 (1999): 132–54.

9 R.C. Allen, "The Two English Agricultural Revolutions, 1450–1850," in *Land, Labour and Livestock*, ed. B.M.S. Campbell and M. Overton (Manchester, New York: St. Martin's Press, 1991), 500.

10 R.L. Hopcroft, *Regions, Institutions, and Agrarian Change in European History* (Ann Arbor: University of Michigan Press, 1999), 272.

11 W.H. Adams, *The French Garden 1500–1800* (New York: Braziller, 1979), 88.

12 Andrew Brink, *English Picturesque and Dutch Landscape Prints of the Seventeenth Century* (Guelph: Macdonald Stewart Art Centre, 2004), 11–31.

13 P.F. Brandon, "Diffusion of designed landscapes," in *Change in the Countryside: Essays on Rural England, 1500–1900*, ed. H.S.A. Fox and R.A. Butlin (London: Institute of British Geographers, 1979), 165–85.

14 Ibid.

15 J.S. Nickerson, *Homage to Malthus* (Port Washington, New York: Kennikat Press, 1975), 47.

CHAPTER TWO

1 Carol Martin, *A History of Canadian Gardening* (Toronto: McArthur & Company, 2000), 36.

2 Charles Gibson, "The Ingham Indian Collection," from *Books at Iowa* 4 (University of Iowa, April 1966), www.lib.uiowa.edu/spec-coll/Bai/gibson.htm.

3 Martin, *A History of Canadian Gardening*, 1–5.

4 W. Cronon, *Changes in the Land* (New York: Hill and Wang, 1983), 24–5.

5 J. Howison, *Sketches of Upper Canada* (Edinburgh: Oliver and Boyd, 1821), 229–30.

6 Patrick Taylor, *Period Gardens: New Life for Historic Landscapes* (New York: Atlantic Monthly Press, 1991), 127.

7 Cited in Taylor, *Period Gardens*, 125–41.

8 Ibid.

9 D. Thompson, *David Thompson's Narrative of his Explorations in*

Western America 1784–1812 (Toronto: The Champlain Society, 1916; facsimile edition: New York: Greenwood Press), 183.

10 J. Todorovová, ed., *Národni Muzeum Praha: A Guidebook: The Náprestek Museum of Asian, African and American Cultures* (Prague: Národní Muzeum Praha, 2000), 23.

11 Ronald Wright, *A Short History of Progress* (Toronto: House of Anansi Press Inc., 2004), 17.

12 J. Russell Harper, ed., *Paul Kane's Frontier: Including "Wanderings of an Artist among the Indians of North America by Paul Kane"* (Toronto: University of Toronto, 1971), 79.

13 Ibid., 99.

14 W.K. Lamb, *Journals and Letters of Sir Alexander Mackenzie* (Cambridge: Cambridge University Press, 1970), 411.

15 T.P. Slaughter, *Exploring Lewis and Clark: Reflections on Men and Wilderness* (New York: Random House, 2003), 231.

16 F.J. Turner, *Significance of the Frontier in American History* (Ann Arbor: University Microfilms, 1966 [1928]), 227.

17 R.A. Billington, *Land of Savagery, Land of Promise: The European Image of the American Frontier in the Nineteenth Century* (New York: Norton, 1981), 364.

18 F.J. Turner, *Significance of the Frontier*, 290.

19 J.G. Brown, *Essay on the Advantages of Canals to the Farmers of Canada* (Toronto: A.H. Amour, 1850).

20 F.J. Turner, *Rise of the New West* (New York: Harper & bros., *circa* 1906).

21 John Howison, *Sketches of Upper Canada* (Edinburgh: Oliver & Boyd, 1821), 305.

22 E.C. Guillet, *Early Life in Upper Canada* (Toronto: University of Toronto Press, 1933), 100.

23 Ontario Agricultural Commission 1880, *Report of the Commissioners*, Vol. III (Toronto: Ontario Government Publications, 1881), 107–8.

24 Jancis Robinson, *The Oxford Companion to Wine* (New York: Oxford University Press, 1994), 725–8.

25 *Brookgreen Gardens* [catalogue]. (Murrells Inlet, SC: Brookgreen Gardens, 1999), www.brookgreen.org.

26 *Oscar Bluemner: In Retrospect* (Minneapolis: University of Minnesota), 1939.

27 Senate and House of Representatives of the United States of America, *Hatch Act* (USA: Congress, 2 March 1887), www.higher-ed.org/resources/hatch.htm (accessed 15 July 2007).
28 J.S. Nielson, *The Poems of Shaw Neilson*, ed. A.R. Chisholm (Sydney: Angus & Robertson, 1965), 227.
29 T.M. Perry, *Australia's First Frontier: The Spread of Settlement in New South Wales, 1788–1829* (Melbourne: Melbourne University Press, 1963), 162.
30 Ibid., 126.
31 John Lhotsky, *A Journey from Sydney and the Australian Alps* (Hobart: Blubber Head Press, 1979), 280.
32 J.C. Weaver, "Beyond the Fatal Shore: Pastoral Squatting and the Occupation of Australia, 1826 to 1852," *American Historical Review* 101, no. 4 (1996): 981–1007.
33 A.J. Christopher, *South Africa* (London: Longman, 1982), 237.
34 P.N. Limerick, *The Legacy of Conquest: The Unbroken Past of the American West* (New York: Norton, 1988), 396.
35 Craig J. Pearson, H. Zuo, I. Valentine, and M. Unkovich, "Sustainability of Grazing Systems," *International Journal of Agricultural Sustainability* 1 (2003): 95–107.
36 Cited in D. Berry, "Defusing the Salt Time Bomb," *Journal of Agriculture, Western Australia* 38 (1997): 35–41.
37 T. Bonyhardy, *The Colonial Earth* (Melbourne: Miegunyah Press, 2000), 5.
38 M.P. Cowan and R.W. Shenton, "Agrarian doctrines of development," *Journal of Peasant Studies* 25 (1998, originally published by the Province of Canada 1849): 49–76.
39 A.O. Craven, *Soil Exhaustion as a Factor in the Agricultural History of Virginia and Maryland 1606–1860* (Urbana: University of Illinois, 1926).
40 Woody Guthrie, "Dust Storm Disaster" ("The Great Dust Storm") (New York: TRO-Ludlow Music Inc., 1960), www.woodyguthrie.org/Lyrics/Dust_Storm_Disaster.htm (accessed 16 Dec. 2006).
41 R. Douglas Hurt, *American Agriculture: A Brief History* (Ames: Iowa State University Press, 1994), 412.
42 E.J. Zavitz, *Fifty Years of Reforestation in Ontario* (Toronto: Ontario Department of Lands and Forests, 1961), 28.
43 Ibid.
44 M.A. Abu-Zeid and F.Z. El-Shibini, "Egypt's High Aswan Dam," in *International Journal of Water Resources Development* 13 no. 2 (1997): 209–17.

CHAPTER THREE

1 Ingo F. Walther, ed., *Masterpieces of Western Art: A History of Art in 900 Individual Studies from the Gothic to the Present Day* (Köln: Taschen, 2002), 442.

2 A.R. Wilkes, "Arable Farming After the Napoleonic Wars," *Agricultural History Review* 28 (1980): 90–103.

3 George Edwin Fussell, *Landscape Painting and the Agricultural Revolution* (London: Pindar Press, 1984), 65.

4 John Fitzherbert and Sir Anthony Fitzherbert, *Booke of Husbandrye* (London: Iohn Awdely, 1562, reprinted 1923).

5 J.L. Van Zanden, "The first green revolution: the growth of production and productivity in European agriculture," *Economic History Review* 44 (1991): 215–39.

6 J. Russell Manning, *The Illustrated Stock Doctor and Livestock Encyclopedia* (Philadelphia: Hubbard Bros., 1882 [first publication N.D. Thompson & Co., St. Louis, Mo., 1880].

7 Sir J.B. Lawes and Sir J.H. Gilbert, *The Rothamsted Experiments* (Edinburgh: W. Blackwood, 1895), 361.

8 Gregor Mendel, "Versuche über Plflanzenhybriden," *Verhandlungen des naturforschenden Vereines in Brünn*, Bd. IV für das Jahr 1865, Abhandlungen, 3–47. Translation: "Experiments in Plant Hybridization," *Journal of the Royal Horticultural Society*, trans. William Bateson, 1901.

9 Rachel Carson, *Silent Spring* (Cambridge: Riverside Press, 1962), 368.

10 V.M. Stern, R.F. Smith, R. van den Bosch, and K.S. Hagen, "The integrated control concept," *Hilgardia* 1959 (Oct.): 81–101.

11 Food and Agriculture Organization Panel of Experts on Pest Control, *Report of the First Session of the FAO Panel of Experts on Integrated Pest Control* (Rome: FAO, 1968).

12 R.F. Chandler to R. Bayfield, 12 May 1958. Sleepy Hollow, NY: Rockefeller Archive Center.

13 Memorandum of 8 October 1958, in Robert F. Chandler, *An Adventure in Applied Science: A History of the International Rice Research Institute,* (Los Baños, Laguna, Philippines: International Rice Research Institute, 1982), Appendix 1, www.irri.org/publications/chandler/pdfs/Appendices.pdf.

14 Rockefeller Foundation, "Memorandum of understanding between government of Mexico and Rockefeller Foundation," 14 Sept 1965 (Sleepy Hollow, NY: Rockefeller Archive Center).

15 Rockefeller Foundation, "Partnership agreement between the government of Mexico and Rockefeller Foundation," 12 April 1966. Sleepy Hollow, NY: Rockefeller Archive Center.

16 Rockefeller Foundation, "President's ten-year review and annual report 1971" (Sleepy Hollow, NY: Rockefeller Foundation, 1971), 30.

17 R.E. Everson and D. Gollin, "Assessing the impact of the green revolution, 1960–2000," *Science* 300 (2003): 758–62.

18 Ibid.

19 J.A. Paine et al, "Improving the nutritional value of Golden Rice through increased pro-vitamin A content," *Nature Biotechnology* 23 (2005): 482–7.

20. Everson and Gollin, "Assessing the impact of the green revolution."

CHAPTER FOUR

1 J.N. Pretty, C. Brett, D. Gee, R.E. Hine, C.F. Mason, J.I.L. Morison, et al., "An assessment of the total external costs of UK agriculture," *Agricultural Systems* 65, no. 2 (2000): 113–36.

2 Reinhard Reitzenstein, *escarpment, valley, desert* [exhibition catalogue] (Hamilton: Art Gallery of Hamilton, 2002), 88.

3 Pretty et al., "Assessment of the total external costs."

4 Department for Environment, Food and Rural Affairs (DEFRA), Origin of the UK Foot and Mouth Disease epidemic in 2001 (June 2002) www.defra.gov.uk/footandmouth/pdf/fmdorigins1.pdf (accessed 30 Oct. 2006).

5 D. Campbell and R. Lee, "'Carnage by computer': the blackboard economics of the 2001 foot and mouth epidemic," Centre for Business Relationships, Accountability, Sustainability and Society & Cardiff Law School (2001), www.fmd.brass.cf.ac.uklcarnagebycomputerDCBL.pdf (accessed Sept. 2004).

6 "It was a hell of a mess," BBC News: Politics (22 July 2002), http://news.bbc.co.uk/1/hi/uk_politics/2144145.stm (accessed 26 June 2007).

7 Ibid.

8 A. Blake, M. T. Sinclair, and G. Sugiyarto, "Quantifying the impact of foot and mouth disease on tourism and the UK economy," *Tourism Economics* 9, no. 3 (2003): 449–65.

9 G. Horn, M. Bobrow, M. Bruce, M. Goedert, A. McLean, and J. Webster, *Review of the origin of BSE* (London: Department of Environment, Food and Rural Affairs, 5 July 5), www.defra.gov.uklanimalhlbse/publications/bseorigin.pdf (accessed 3 Nov. 2006).

10 Sean Callahan, ed., *The Photographs of Margaret Bourke-White* (Greenwich, CN: New York Graphic Society Ltd., 1972), 38.

11 David Waltner-Toews, "A Bill from the Power Company," in *The Fat Lady Struck Dumb* (London, Ontario: Brick Books, 2000), 25.

12 Annie Proulx, *That Old Ace in the Hole* (New York: Scribner, 2002), 359.

13 Linda M. Lobao, Industrialized farming and its relationship to community well-being: report prepared for the State of South Dakota, Office of the Attorney General (Columbus, Ohio: Ohio State University, 2000), www.agribusinessaccountability .org/pdfs/270 _lndustrialized%20Farm ing.pdf (accessed Jan. 2007).

14 S. Wing, D. Cole, and G. Grant, "Environmental injustice in North Carolina's hog industry," *Environmental Health Perspectives* 103, no. 3, (2000): 225–31.

15 William J. Weida, Comments on the permit materials submitted by DGH and the Taiwan Sugar Corporation concerning economic development and a farrow-to-finish hog operation in the county of Flagstaff, Alberta, Canada (New York: Global Resource Action Centre for the Environment, 2000), www.factoryfarm.org/docs/hardestyappealmaster1000.doc (accessed 16 Dec. 2006).

16 J.K. Galbraith, *The Economics of Innocent Fraud: Truth for Our Time* (Boston: Houghton Mifflin, 2004): 62.

CHAPTER FIVE

1 Natural Land and Water Resources, Australia, *Australian Dryland Salinity Assessment 2000* (Canberra: Land and Water Australia and Goanna Press, 2001), 162.

2 Cosmo Cantuar [Archbishop of Canterbury], Cardinal Hinsley, Walter. H. Armstrong, and William Ebor, "Foundations of Peace," *The Times* (London), 21 Dec. 1940, 5–6.

3 Aldo Leopold, *A Sand County Almanac*, and sketches here and there (New York: Oxford University Press, 1987), 81.

4 The Wilderness Society, "Who we are," Wilderness Society website, 2006) www.wilderness.org/AboutUs/who.cfm (accessed 3 Nov. 2006)

5 Albert Howard, *An Agricultural Testament* (London: Oxford University Press, 1940).

6 Lord Northbourne, *Look to the Land* (London: Dent, 1940).

7 Albert Howard, *The Soil and Health: A Study of Organic Agriculture* (New York: Schocken Books, *circa* 1947, reprinted 1972).

8 Lady Evelyn Balfour, *The Living Soil: Evidence of the Importance to Human Health of Soil Vitality, with Special Reference to Post-War Planning* (London: Faber and Faber, 1943).

9 Soil Association, "A Brief History of the Soil Association," Soil Association website, www.soilassociation.org/web/sa/saweb.nsf/Library?OpenForm&Cat=_History (accessed 18 May 2007).

10 R. Wolf, *Organic Farming: Yesterday's and Tomorrow's Agriculture* (Emmaus, Pa.: Rodale Press, 1977), 344.

11 J.C. Zadoks, *Development of Farming Systems – Evaluation of the Five Year Period 1980–1984* (Netherlands: Centrum voor Landbouwpublikaties en Landbouwdocumentatie, 1989), 90.

12 Carol Martin, *A History of Canadian Gardening* (Toronto: McArthur & Company, 2000), 133–41.

13 Albert Schweitzer, *Peace or Atomic War?* (New York: Henry Holt and Company, 1958), 58.

14 Donella H. Meadows, Dennis L. Meadows, Jorgen Randers, and William W. Behrens III, *The Limits to Growth: A Report for the Club of Rome's Project on the Predicament of Mankind* (New York: Universe Books, 1972), 205.

15 Peter Checkland, *Systems Thinking, Systems Practice* (New York: J. Wiley, 1981), 330; C.J. Pearson and R.L. Ison, *Agronomy of Grassland Systems*, 2nd ed. (Cambridge: Cambridge University Press, 1997), 222.

16 United Nations General Assembly, Rio Declaration on Environment and Development, *Report on the United Nations Conference on Environment and Development* (Rio de Janeiro: United Nations, 3–14 June 1992), www.unep.org/Documents.Multilingual/Default.asp?DocumentID=78&ArticleID=1163 (accessed 2007 May 12).

17 D.J. Rapport, W.S. Fyfe, R. Costanza, J. Spiegel, A. Yassi, G.M. Bohm, et al., "Ecosystem health: definitions, assessment, and case studies," in *Our*

Fragile World: Challenges and Opportunities for Sustainable Development, ed. M.K. Tolba (Oxford: EOLSS Publishers, 2001): 21–42.

18 D.J. Rapport and W.G. Whitford, "How ecosystems respond to stress," *BioScience* 49 (1999): 193–203.

19 David Skole and Compton Tucker, "Tropical deforestation and habitat fragmentation in the Amazon," *Science* 260, no. 5116 (1993): 1905–10.

20 Commission of the European Communities, *The Development and Future of the Common Agricultural Policy: Proposals of the Commission of the European Communities*, COM (91): 258.

21 Craig J. Pearson to B. Fisher and T. Podbury of Australian Bureau of Agricultural and Resource Economics, Canberra, 12 Jan. 2004, email correspondence.

22 Ibid.

23 European Commission, "Council Directive 9/676/EEC," *Official Journal* L375, 1991, 1–8.

24 J. Bouma, B.J. van Alphen, and J.J. Stoorvogel, "Fine tuning water quality regulations in agriculture to soil differences," *Environmental Science and Policy* 5, no. (2002): 113–20; J. Bouma, "Land quality indicators of sustainable land management across scales," *Agriculture Ecosystems and Environment* 88, no. 2 (2002): 129–36.

25 J. Hanson, "Nitrate balances in agriculture," *Statistics in Focus* [European Commission: Eurostat] 8, no. 16 (2000): 1–8.

26 Marcel de Wit, Horst Behrendt, Giuseppe Bendoricchio, Wladimir Bleuten, Pauline van Gaans, "The contribution of agriculture to nutrient pollution in three European rivers, with reference to the European Nitrtates Directive," *European Water Management* [electronic journal] 2 (2002), www.ewaonline.de/journal/online.htm (accessed 2007 May 12).

27 United Nations General Assembly, Rio Declaration on Environment and Development, cited in n. 17, above.

28 World Economic Forum, *2001 Environmental Sustainability Index: An Initiative of the Global Leaders of Tomorrow Environment Task Force*, Davos, Switzerland (2001), http://sedac.ciesin.org/es/esi/ESI_01a.pdf (accessed 2007 May 12).

29 G.L. Brinkman, An Update of Agri-food Research and Technology Transfer Capacity in Canada Through the 1990s. Report submitted to Ontario Ministry of Agriculture and Food, May 2001.

30 Joan E. Kirner, *Landcare: Its Origins* (Melbourne: Landcare Australia, 2000).
31 N. Röling, "Platforms for decision-making about ecosystems," in *The Future of the Land*, ed. J.O. Fresco et al. (New York: John Wiley & Sons, 1994), 385–93.
32 Kirner, *Landcare: Its Origins.*
33 J. Bouma, "Land quality indicators of sustainable land management" *Agriculture Ecosystems and Environment* 88, no. 2 (2002): 129–36.
34 Mathis Wackernagel and William E. Rees, *Our Ecological Footprint: Reducing Human Impact on the Earth* (Gabriola Island, BC: New Society Publishers, 1996), 160; Jason Venetoulis, Dahlia Chazan, and Christopher Gaudet, "Ecological footprint of nations," in *Redefining Progress* (Oakland, CA: Redefining Progress, 2004), 14.
35 S. Frey, D. Harrison, and E. Billett, "Environmental assessment of electronic products using LCA and ecological footprint" in *Electronics Goes Green 2000*, Berlin, Germany, 11–13 Sept. 2000 (Egham, UK: Brunel University, Cleaner Electronics Research Group, Runnymede Campus, 2000), 253–8.
36 Sarah Quinton, ed., *Stitching Women's Lives: Sujuni and Khatwa from Bihar, India* (Toronto: Textile Museum of Canada, 2000), 12–16.

CHAPTER SIX

1 Crescencia Maurer, Suzanne Ehlers, and Andrew Buchman, *Aligning Commitments: Public Participation, International Decision-Making and the Environment* (Washington, DC: World Resources Institute, 2003), www.wri.org/governance/pubs_description.cfm?pid=3796 (accessed 6 June 2007).
2 Mathis Wackernagel and William E. Rees, *Our Ecological Footprint: Reducing Human Impact on the Earth* (Gabriola Island, BC: New Society Publishers, 1996).
3 Gutteridge, Haskins & Davey. "An investigation of nutrient pollution in the Murray-Darling river system," *Report for the Murray-Darling Basin Commission* (Canberra: GHD, 1992).
4 Department for Environment, Food and Rural Affairs (DEFRA), *Consumer Products: The European Ecolabel* (11 Jan. 11, 2005), www.defra.gov.uk/environment/consumerprod/ecolabel/intro.htm (accessed 30 Oct. 2006).

5 J. Fernandez-Cornejo and M. Caswell, "The first decade of genetically engineered crops in the United States." *USDA Economic Information Bulletin* 2 (2006): 30.
6 Keith J. Betteridge, "A history of farm animal embryo transfer and some associated techniques," *Animal Reproductive Science* 79 (2003): 203–44.
7 I. Coxhead and D. Southgate, "Economy-wide sources of agricultural expansion in developing countries," *International Journal of Agricultural Resources, Governance and Ecology* 1 (2000): 68–76.
8 Gretchen C. Daily, "Introduction: What are ecosystem services?" *Nature's Services: Societal Dependence on Natural Ecosystems* (Washington, DC: Island Press, 1997), 1–10.
9 Maurer et al, *Aligning Commitments.*
10 L.R. Oldeman, "The Global Extent of Soil Degradation," in *Soil Resilience and Sustainable Land Use: Proceedings of a Symposium Held in Budapest, 28 September – 2 October 1992*, ed. D.J. Greenland and I. Szabolcs (Wallingford, UK: CAB International, 1994), 561.
11 Ingeborg Boyens, *Another Season's Promise: Hope and Despair in Canada's Farm Country,* (Toronto: Viking/Penguin, 2001).
12 National Land and Water Resources, *Australian Dryland Salinity Assessment 2000: Extent, impacts, processes, monitoring and management options.* (Braddon, Australia, Jan. 2001), http://audit.ea.gov.au/ANRA/land/docs/national/Salinity_Contents.html (accessed 6 June 2007).
13 Maurer et al, *Aligning Commitments.*
14 Australian Government, *State of the Environment 2001 (SoE 2001)* (Canberra: Department of the Environment and Water Resources, 1996–2007), www.deh.gov.au/soe/2001/index.html (accessed Oct. 2006).
15 Ibid.
16 Nancy Hofmann, *Urban Consumption of Agricultural Land*, 3 (2) (Ottawa: Statistics Canada, 5 Sept. 2001) www.statcan.ca/bsolc/english/bsolc?catno=21-006-X2001002 (accessed Jan. 2007).
17 M. Grubb et al., *The Kyoto Protocol: A Guide and Assessment* (London: Royal Institute of International Affairs, Energy and Environmental Programme, 1999), 342.
18 World Resource Institute, "WRI findings underscore growing crisis of confidence in international institutions" [news release], (Washington, DC: World Resource Institute, June 5, 2003) http://governance.wri.org/news release_text.cfm?NewsReleaseID=247 (accessed 7 June 2007).

CHAPTER SEVEN

1 J. Jiggins and N. Röling, "Adaptive management: potential and limitations for ecological governance," *International Journal of Agricultural Resources*, 1 (2000): 28–41.

2 D. Ranney, "Mobile capital and economic development planning," *Journal of Planning Education and Research* 20 (2001): 281–92.

3 Joseph E. Stiglitz, *Globalization and Its Discontents* (New York: W.W. Norton & Company, 2002).

4 Margaret A. Somerville and David J. Rapport, editors. *Transdisciplinarity: Recreating Integrated Knowledge* (Montreal: McGill-Queen's University Press, 2000), 272.

5 J. Peterson, Fiona Cornwell and C.J. Pearson, *Chain Stocktake of Some Australian Agricultural and Fishing Industries* (Canberra: Bureau of Rural Sciences, 2000), 15.

6 Andrew Brink, *English Picturesque and Dutch Landscape Prints of the Seventeenth Century* (Guelph: Macdonald Stewart Art Centre, 2004), 11–31.

7 L.O. Mearns, "Issues in the impacts of climate variability and change on agriculture: applications to the southeastern United States," *Climatic Change* 60 (2003): 1–6.

8 Craig Pearson, "Regenerative or semi-closed systems: a priority for twenty-first century agriculture," *BioScience* 57 (2007): 409–18.

9 Australia Parliament House of Representatives, Standing Committee on Environment and Heritage, *Inquiry into Public Good Conservation* (Canberra: Commonwealth of Australia, 2002), www.aph.gov.au/house/committee/environ/pubgood/report/contents.htm.

10 Craig Canine, "Building a Better Banana," *Smithsonian Magazine* (October 2005), http://www.smithsonianmagazine.com/issues/2005/october/banana.htm (accessed 2007 September 26).

11 Allen Consulting, "Repairing the Country. Leveraging Private Investment: A Discussion Paper," prepared for the Business Leaders Roundtable (Canberra/Sydney: Allen Consulting Group, 2001): 12, www.acfonline.org-au/uploads/res_private_investment.pdf.

12 G.M. Woodwell, "The functional integrity of normally forested landscapes: A proposal for an index of environmental capital," *Proceedings of the National Academy of Sciences* 99, (2002): 13600–5.

13 R. Thackway and I.D. Cresswell, "A bioregional framework for planning the national system of protected areas in Australia," *Natural Areas Journal* (1997): 241–7.
14 Kristen Helmore and Naresh Singh, *Sustainable Livelihoods: Building on the Wealth of the Poor* (Bloomfield, Conn,: Kumarian Press, 2001), 128.
15 The National Agricultural Centre, LEAF (Linking Environment and Farming) (Warwickshire: The National Agricultural Centre, 2003), www.leafuk.org/leaf.
16 J.L. Nassauer, R.C. Corry, and R.M. Cruse, "Alternative future landscape scenarios: a means to consider agricultural policy," *Journal of Soil and Water Conservation* 57 (2002): 44–53.
17 D.J. Brunkhorst, *Bioregional Planning: Resource Management Beyond the New Millennium* (London: Routledge, 2001), 159.
18 T.S. Eliot, "Burnt Norton," *Four Quartets*. In: *The Complete Poems and Plays of T.S. Eliot* (London: Faber & Faber, 1969).

BIBLIOGRAPHY

Abu-Zeid, M.A., and F.Z. El-Shibini. "Egypt's High Aswan Dam." *International Journal of Water Resources Development* 13, no. 2 (1997): 209–17.

Adams, W.H. *The French Garden 1500–1800*. New York: Braziller, 1979.

Ainslie, Patricia. *Images of the Land: Canadian Book Prints 1919–1945*, 2nd ed. Calgary: Glenbow Museum, 1985.

Allen Consulting Group. *Repairing the Country. Leveraging Private Investment: A Discussion Paper, Prepared for the Business Leaders Roundtable*. Canberra/Sydney: ACG, 2001.

Allen, R.C. "The Two English Agricultural Revolutions, 1450–1850." In *Land, Labour and Livestock*, edited by B.M.S. Campbell and M. Overton. Manchester, New York: St. Martin's Press, 1991.

Australian Government. *State of the Environment 2001 (SoE 2001)*. Canberra: Department of the Environment and Water Resources, 1996–2007. www.deh.gov.au/soe/2001/index.html.

Australia Parliament House of Representatives. *Standing Committee on Environment and Heritage: Inquiry into Public Good Conservation*. Canberra: Commonwealth of Australia, June 20, 2002. www.aph.gov.au/house/committee/environ/pubgood/report/contents.htm.

Baker, Steve. *Picturing the Beast: Animals, Identity and Representation.* Chicago: University of Illinois Press, 2001.

Balázs, György, ed. *The Abolition of Serfdom and its Impact on Rural Culture.* Budapest/Syentendre: Museum of Hungarian Agriculture, Hungarian Open-Air Museum, 1998.

Balfour, Lady Evelyn. *The Living Soil: Evidence of the Importance to Human Health of Soil Vitality, with Special Reference to Post-War Planning.* London: Faber & Faber, 1943.

Beall, Karen F. and David Kiehl. *Graphic Excursions, American Prints in Black and White, 1900–1950: Selection from the Collection of Reba and Dave Williams.* New York: David R. Godine Publisher, Ltd., 1991.

Bell, Michael. *Painters in a New Land. From Annapolis Royal to the Klondike.* Toronto: McClelland and Stewart, 1973.

Berger, John. *Pig Earth.* New York: Random House, 1992.

Berlo, Janet C., and Ruth B. Phillips. *Native North American Art.* New York: Harvard University Press, 1998.

Berry D. "Defusing the Salt Time Bomb." *Journal of Agriculture – Western Australia* 38 (1997): 35–41.

Betteridge, Keith J. A history of farm animal embryo transfer and some associated techniques. *Animal Reproductive Science* 79 (2003): 203–44.

Billington, R.A. *Land of Savagery, Land of Promise: The European Image of the American Frontier in the Nineteenth Century.* New York: Norton, 1981.

Blake, A., M.T. Sinclair, and G. Sugiyarto. "Quantifying the impact of foot and mouth disease on tourism and the UK economy." *Tourism Economics* 9, no. 4 (2003): 449–65.

Bonyhardy, T. *The Colonial Earth.* Melbourne: Miegunyah Press, 2000.

Bouma, J. "Land quality indicators of sustainable land management across scales." *Agriculture Ecosystems and Environment* 88, no. 2 (2002): 129–36.

Bouma, J., B.J. van Alphen, and J.J. Stoorvogel. "Fine tuning water quality regulations in agriculture to soil differences." *Environmental Science and Policy* 5, no. 2 (2002): 113–20.

Boyens, Ingeborg. *Another Season's Promise: Hope and Despair in Canada's Farm Country.* Toronto: Viking/Penguin, 2001.

Brandon, P.F. "Diffusion of designed landscapes." In *Change in the Countryside: Essays on Rural England, 1500–1900.* Edited by H.S.A. Fox and R.A. Butlin. London: Institute of British Geographers, 1979.

British Broadcasting Corporation. *"It was a hell of a mess." BBC News: Politics*. London: BBC, Monday, July 22, 2002. http://news.bbc.co.uk/1/hi/uk_politics/2144145.stm.

Brink, Andrew. *English Picturesque and Dutch Landscape Prints of the Seventeenth Century*. Guelph: Macdonald Stewart Art Centre, 2004.

Brinkman, G.L. An Update of Agri-food Research and Technology Transfer Capacity in Canada Through the 1990s. Report submitted to Ontario Ministry of Agriculture and Food, May 2001.

Brookgreen Gardens. *Brookgreen Gardens* [catalogue]. Murrells Inlet, South Carolina: Brookgreen Gardens, 1999. www.brookgreen.org.

Brown, J.G. *Essay on the Advantages of Canals to the Farmers of Canada*. Toronto: A.H. Amour, 1850.

Brunkhorst, D.J. *Bioregional Planning: Resource Management Beyond the New Millennium*. Routledge: London, 2001.

Brydon, Anne, Shirley Madill, and Joanne Sloan, editors. *Eleanor Bond: Social Centres*. Winnipeg: Winnipeg Art Gallery, 1993.

Butlin, R.A. "The Enclosure of Open Fields and Extinction of Common Rights in England, c. 1600–1750: A Review." In *Change in the Countryside: Essays on Rural England, 1500–1900*, H.S.A. Fox and R.A. Butlin, editors. London: Institute of British Geographers, 1979: 65–82.

Callahan, Sean, editor. *The Photographs of Margaret Bourke-White*. Greenwich, CN: New York Graphic Society, 1972.

Camille, Michael. *Gothic Art: Glorious Visions*. New York: Harry N. Abrams, 1996.

Campbell, D., and R. Lee. "Carnage by Computer: The Blackboard Economics of the 2001 Foot and Mouth Epidemic." *The Centre for Business Relationships, Accountability, Sustanability and Society & Cardiff Law School*. www.fmd.brass.cf.ac.uk/carnagebycomputerDCBL.pdf.

Campbell, Nancy. *Frankenstein: Explorations in Manipulation and Surrationality*. Guelph: Macdonald Stewart Art Centre, 1994.

Canine, Craig. "Building a Better Banana," *Smithsonian Magazine* (October 2005), http://www.smithsonianmagazine.com/issues/2005/october/banana.htm.

Cantuar, Cosmo [Archbishop of Canterbury], Cardinal Hinsley, Walter. H. Armstrong, and William Ebor. *Foundations of Peace. The Times* (London), 21 Dec. 1940: 5–6.

Carson, R. *Silent Spring*. Cambridge: Riverside Press, 1962.

Cate, Phillip D. *Circa 1800: The Beginnings of Modern Printmaking 1775–1835*. Washington: Rutgers, 1981.

Chandler, R.F. Correspondence with *R. Bayfield* (May 12 1958). Sleepy Hollow, New York: Rockefeller Archive Center.

Chandler, Robert F. *An Adventure in Applied Science: A History of the International Rice Research Institute*. Los Baños, Laguna, Philippines: International Rice Research Institute, 1982, Appendix 1. www.irri.org/publications/chandler/pdfs/Appendices.pdf.

Checkland, P.B. *Systems Thinking, Systems Practice*. New York: J. Wiley, 1981.

Christopher, A.J. *South Africa*. London: Longman, 1982.

Clark, Kenneth. *Landscape into Art*. Boston: Beacon Press, 1961.

Clarke, Philip, Brain Jackman and Derrik Merce, editors. *The Sunday Times Book of the Countryside: Including One Thousand Days Out in Great Britain and Ireland*. Great Britain: Macdonald Futura Publishers, 1980 [reissue].

Colbert, Judith A. *The Achievement and the Challenge: the University of Guelph*. Guelph: University of Guelph, 1989.

Commission of the European Communities. *The Development and Future of the Common Agricultural Policy: Proposals of the Commission of the European Communities*. COM (91), 258.

Columella, Lucius Junius Moderatus. *On Agriculture and Trees*, translated by E.S. Forster and Edward H. Heffner. Cambridge, MA: Harvard University Press, 1954–55.

Cowan, M.P., and R.W. Shenton. "Agrarian doctrines of development." *Journal of Peasant Studies* 25 (1998 [originally published by Province of Canada, 1849]): 49–76.

Coxhead, I., and D. Southgate. "Economy-wide sources of agricultural expansion in developing countries." *International Journal of Agricultural Resources, Governance & Ecology* (2000): 68–76.

Craig-Smith, Stephen J., Craig J. Pearson and Juliet C. Middleton, *Gatton College: 100 Years of Science with Practice*. Queensland: University of Queensland, 1996.

Craven, A.O. *Soil Exhaustion as a Factor in the Agricultural History of Virginia and Maryland 1606–1860*. Urbana: University of Illinois, 1926.

Cronon, W. *Changes in the Land*. New York: Hill and Wang, 1983.

Daily, G.C. "Introduction: What are Ecosystem Services?" In *Nature's Services: Societal Dependence on Natural Ecosystems.* Washington, DC: Island Press, 1997.

de Wit, Marcel, Horst Behrendt, Giuseppe Bendoricchio, Wladimir Bleuten, Pauline van Gaans. "The contribution of agriculture to nutrient pollution in three European rivers, with reference to the European Nitrtates Directive." *European Water Management* [electronic journal] 2 (2002), www.ewaonline.de/journal/online.htm.

Department for Environment, Food and Rural Affairs (DEFRA). *Origin of the UK Foot and Mouth Disease epidemic in 2001* in *Department of the Environment, Food and Rural Affairs.* DEFRA website, June 2002, www.defra.gov.uk/footandmouth/pdf/fmdorigins1.pdf.

Department for Environment, Food and Rural Affairs (DEFRA). "Consumer Products: The European Ecolabel." *Department for Environment, Food and Rural Affairs.* DEFRA website, 11 Jan. 2005, www.defra.gov.uk/environment/consumerprod/ecolabel/intro.htm.

Dewdney, Selwyn. *Sacred Scrolls of the Southern Ojibway.* Toronto: University of Toronto Press, 1975.

Diamond, Jared. *Collapse: How Societies Choose to Fall or Succeed.* New York: Viking, 2005.

Durrenberger, Paul E., and M. Thu Kendall. "The expansion of large scale hog farming in Iowa: the applicability of Goldschmidt's findings fifty years later." *Human Organization* 55, no. 4 (1996): 409–15.

Edelstein, Debra. *Graphic Excursions: American Prints in Black and White 1900–1950: Selections from the Collection of Reba and Dave Williams.* New York: The American Federation of Arts, 2001.

Eliot, T.S. *The Complete Poems and Plays of T.S. Eliot.* London, Faber & Faber, 1969.

European Commission. "Council Directive 9/676/EEC." *Official Journal,* L375, 1991: 1–8.

Evans, Walker. *American Photographs.* New York: East River Press, 1975.

Everson, R.E., and D. Gollin. "Assessing the impact of the green revolution, 1960–2000." *Science* 300 (2003): 758–62.

Falvey, Emily. *Dead Nature/La vie Immobile: Rachel Echenberg, Louis Joncas & Kelly Mark.* Ottawa: Ottawa Art Gallery, 2005.

Fernandez-Cornejo, J., and M. Caswell. "The first decade of genetically engineered crops in the United States." *USDA Economic Information Bulletin* 2 (2006): 30.

Fischer, Barbara, ed. *Foodculture: Tasting Identities and Geographies in Art.* Toronto: YYZ Books, 1999.

Fitzherbert, Sir John and Sir Anthony Fitzherbert. *Booke of Husbandrye.* London: Iohn Awdely, 1562, reprinted 1923.

Food and Agriculture Organization Panel of Experts on Pest Control. *Report of the First Session of the FAO Panel of Experts on Integrated Pest Control* (Rome: FAO, 1968).

Forster, E.S. and Edward H. Heffner trans. *Lucius Junius Moderatus Columella On Agriculture.* Cambridge, Massachusetts: Harvard University Press, 1954–55.

Frey, S., D. Harrison, and E. Billett. "Environmental assessment of electronic products using LCA and ecological footprint." In *Electronics Goes Green 2000*, Berlin, Germany, 11–13 Sept. 2000 (Egham, UK: Brunel University, Cleaner Electronics Research Group, Runnymede Campus, 2000), 253–8.

Friedlander, Lee. *The Lee Friedlander Photographs.* New York: Haywire Press, 1978.

Fussell, G.E. *Landscape Painting and the Agricultural Revolution.* London: Pindar Press, 1984.

Galbraith, J.K. *The Economics of Innocent Fraud: Truth for Our Time.* Boston: Houghton Mifflin, 2004.

Gibson, Charles. "The Ingham Indian Collection," from *Books at Iowa* 4. University of Iowa, April 1966. www.lib.uiowa.edu/spec-coll/Bai/gibson.htm.

Godine, David R., editor. *200 Years of American Sculpture.* New York: Whitney Museum of American Art, 1976.

Grubb, M., et al. *The Kyoto Protocol: A Guide and Assessment.* London: Royal Institute of International Affairs, Energy and Environmental Programme, 1999.

Guaita, Ovidio. *Terrestrial Paradise.* New York: The Monacelli Press Inc, 1999.

Guillet, E.C. *Early Life in Upper Canada.* Toronto: University of Toronto Press, 1933.

Guthrie, Woody. "Dust Storm Disaster" ("The Great Dust Storm"). New York: TRO-Ludlow Music Inc, 1960, www.woodyguthrie.org/Lyrics/Dust_Storm_Disaster.htm.

Gutteridge, Haskins & Davey. "An investigation of nutrient pollution in the Murray-Darling river system." *Report for the Murray-Darling Basin Commission.* Canberra: GHD, 1992.

Hanna, Martha. *Orest Semchishen: In Plain View.* Ottawa: Canadian Museum of Contemporary Photography, 1994.

Hanson, J. "Nitrate balances in agriculture." *Statistics in Focus* 8, no. 16 (2000) [European Commission: Eurostat].

Harper, J. Russell, editor. *Paul Kane's Frontier: Including "Wanderings of an Artist among the Indians of North America by Paul Kane."* Toronto: University of Toronto, 1971.

Harper, J. Russell, ed. *A People's Art: Primitive, Naïve, Provincial, and Folk Painting in Canada.* Toronto: University of Toronto Press, 1974.

Hartt, Frederick. *History of Italian Renaissance Art. Painting, Sculpture, Architecture,* 4th ed. New York: Harry N. Abrams Inc., 1994.

Hassrick, Peter. *The Way West: Art of Frontier America.* New York: Abradale Press/Harry N. Abrams Inc., 1983.

Helmore, K., and N. Singh. *Sustainable Livelihoods: Building on the Wealth of the Poor.* Bloomfield, Connecticut: Kumarian Press, 2001.

Hofmann, Nancy. *Urban Consumption of Agricultural Land* 3(2). Ottawa: Statistics Canada, September 5, 2001, www.statcan.ca/bsolc/english/bsolc?catno=21-006-X2001002.

Hopcroft, R.L. *Regions, Institutions, and Agrarian Change in European History.* Ann Arbor: University of Michigan Press, 1999.

Horn, G., M. Bobrow, M. Bruce, M. Goedert, A. McLean, and J. Webster. *Review of the origin of BSE.* London: Department of Environment, Food and Rural Affairs, 5 July 2001: 4, www.defra.gov.uklanimalhl bse/publications/bseorigin.pdf.

Howard, A. *An Agricultural Testament.* London: Oxford University Press, 1940.

Howard, A. *The Soil and Health: a study of organic agriculture.* New York: Schocken Books, *circa* 1947, reprint 1972.

Howison, J. *Sketches of Upper Canada.* Edinburgh: Oliver and Boyd, 1921.

Hubbard, R.H., ed. *Canadian Landscape Painting 1670–1930: The Artist and the Land*. Madison: University of Wisconsin, 1973.

Hurt, R.D. *American Agriculture: A Brief History*. Ames: Iowa State University Press, 1994.

Ivins Jr., William M. *Prints and Visual Communication*. Cambridge: MIT Press, 1969.

Jenkner, Ingrid. *The Ontario Heritage Foundation: Stewart and Letty Bennett Collection*. Guelph: Macdonald Stewart Art Centre, 1984.

Jiggins, J., and Röling, N. "Adaptive management: potential and limitations for ecological governance." *International Journal of Agricultural Resources* 1 (2000): 28–41.

Jones, Robert L. *History of Agriculture in Ontario: 1613–1880*. Toronto: University of Toronto Press, 1946.

Joseph Jr., Alvin M. *The Artist was a Young Man: The Life Story of Peter Rindisbacher*. Fort Worth, Texas: Amon Carter Museum of Western Art, 1970.

Kimbrell, Andrew, ed. *The Fatal Harvest Reader: The Tragedy of Industrial Agriculture*. Washington: Island Press, 2002.

Kirner, Joan E. *Landcare: Its Origins*. Melbourne: Landcare Australia, 2000.

Kirsten, Lincoln. *Walker Evans: American Photographs*. New York: East River Press, 1975.

Klein, H. Arthur, ed. *Graphic Worlds of Peter Brueghel the Elder*. New York: Dover Publications Inc., 1963.

Land and Water Australia. *Australian Dryland Salinity Assessment 2000: Extent, impact, processes, monitoring and management options*. Canberra: Commonwealth of Australia, 2001, http://audit.ea.gov.au/ANRA/land/docs/national/Salinity_Contents.html (accessed December 16, 2006)

Lamb, W. K. *Journals and Letters of Sir Alexander Mackenzie*. Cambridge: Cambridge University Press, 1970.

Langford, Martha. *Beau*. Ottawa: Canadian Museum of Contemporary Photography, 1988.

Lawes, Sir J.B. and J.H. Gilbert. *The Rothamsted Experiments*. Edinburgh: W. Blackwood, 1895.

Lear, Edward. *Book of Nonsense*. New York: Metropolitan Museum of Art, 1980.

Leopold, A. *A Sand County Almanac, and Sketches Here and There*. New York: Oxford University Press, 1987.

Lhotsky, J. *A Journey from Sydney and the Australian Alps*. Hobart: Blubber Head Press, 1979.

Limerick, P.N. *The Legacy of Conquest: the Unbroken Past of the American West*. New York: Norton. 1988.

Little, J.I. "Agricultural Improvement and Highland Clearance: The Isle of Arran, 1828–29." *Scottish Economic and Social History* 19 (1999): 132–54.

Lobao, Linda M. *Industrialized Farming and its Relationship to Community Well-being*. Columbus, OH: Ohio State University, 2000, www.agribusinessaccountability.org/pdfs/270_Industrialized%20Farming.pdf.

Logan, Robert A., ed. *Great Patriotic War: A Collection of World War II Soviet Propaganda Posters*. Guelph: University of Guelph, 1984.

Love, Karen. *Weathervane L'air du Temps*. Montreal: The Ottawa Art Gallery/Oakville Galleries, 2005.

MacLennan, Hugh. *Contemporary Canadian Photography: From the Collection of the National Film Board*. Edmonton: Hurtig Publishers Ltd., 1984.

Mallet, Gina. *Last Chance to Eat: The Fate of Taste in a Fast Food World*. New York: W. W. Norton & Company Inc., 2004.

Manning, Russell J. *The Illustrated Stock Doctor and Livestock Encyclopaedia*. Philadelphia: Hubbard Bros., 1882.

Marchand, Sandra Grant, and Johanne Sloan. *Eleanor Bond*. Montreal: Musée d'art contemporain de Montréal/La Direction, 1998.

Martin, Carol. *A History of Canadian Gardening*. Toronto: McArthur & Company, 2000.

Mathis-Moser, Ursula, Gerard van Bussel, Sybille-Karin Moser, Elke Nowak, and Judith Nasby. *Asingit: Inuit Art from Macdonald Stewart Art Centre*. Edited by Ursula Mathis-Moser and Sybille-Karin Moser. Innsbruck, Austria: Kunstgeschichte der Leopold-Franzens-Universitat Innsbruck, 2002.

Maurer, Crescencia, Suzanne Ehlers, and Andrew Buchman. *Aligning Commitments: Public Participation, International Decision-Making, and the Environment.* Washington DC: World Resources Institute, 2003, www.wri.org/goverance/pubs_description.cfm?pid=3796.

McCulloch, Susan. *Contemporary Aboriginal Art: A Guide to the Rebirth of an Ancient Culture.* Australia: Allen & Unwin, 1999.

Meadows, Donella H., Dennis L. Meadows, Jorgen Randers, and William W. Behrens III. *The Limits to Growth: A Report for the Club of Rome's Project on the Predicament of Mankind.* New York: Universe Books, 1972.

Mearns, L. O. "Issues in the Impacts of Climate Variability and Change on Agriculture: Applications to the southeastern United States." *Climatic Change* 60 (2003): 1–6.

Mehrotra, A.K. *Middle Earth.* New York: Oxford University Press, 1984.

Mendel, Gregor. "Versuche über Plflanzenhybriden." *Verhandlungen des naturforschenden Vereines in Brünn,* Bd. IV für das Jahr 1865, Abhandlungen, 3–47. Translation: "Experiments in Plant Hybridization," *Journal of the Royal Horticultural Society,* trans. William Bateson, 1901.

Murray, Joan. *Kurelek's Vision of Canada.* Oshawa: Robert McLaughlin Gallery, 1982.

Museum of Contemporary Art. *La Frontera/The Border: Art about the Mexico/United States Border Experience.* Mexico: Centro Cultural De La Raza / San Diego: Museum of Contemporary Art, 1993.

Nasby, David. *Permanence and Change: A Rural Ontario Document.* Canada: Herzig-Somerville, 1973.

Nasby, Judith M. *Canadian Inuit Art: Selections from the Collection of Dr. and Mrs. E. Albrecht.* Phoenix: Heard Museum, 2000.

Nasby, Judith M. *Contemporary Inuit Drawings: The Gift Collection of Frederick and Lucy S. Herman.* Williamsburg, Virginia: Muscarelle Museum of Art, College of William and Mary, 1993.

Nasby, Judith M. *Guelph Artists 2002.* Guelph: Macdonald Stewart Art Centre, 2002.

Nasby, Judith M. Inuit Art: 1950–2002, an exhibition at Macdonald Stewart Art Centre. *Inuit Art Quarterly* Summer (2002): 50–52.

Nasby, Judith M. *Qamanittuaq Drawings by Baker Lake Artists from the Collection of Macdonald Stewart Art Centre.* Guelph: Macdonald Stewart Art Centre, 1998.

Nasby, Judith M. "Revealing the Truth of the Artist's Hand: Contemporary Inuit Drawing." *Inuit Art Quarterly* Fall (1994): 4–13.

Nasby, Judith M. *The University of Guelph Art Collection: A Catalogue of Paintings, Drawings, Prints and Sculpture.* Guelph: University of Guelph, 1980.

Nassauer, J.I., R.C. Corry, and R.M. Cruse. "Alternative future landscape scenarios: a means to consider agricultural policy." *Journal of Soil and Water Conservation* 57 (2002): 44–53.

National Agricultural Centre. LEAF [Linking Environment and Farming]. Warwickshire: National Agricultural Centre, 2003, www.leafuk.org/leaf.

National Land and Water Resources. *Australian Dryland Salinity Assessment 2000: Extent, impacts, processes, monitoring and management options.* (Braddon, Australia, Jan. 2001), http://audit.ea.gov.au/ANRA/land/docs/national/Salinity_Contents.html.

Newhall, Beaumont. *The History of Photography from 1839 to the Present Day.* New York: Museum of Modern Art, 1964.

Nielson, J.S. *The Poems of Shaw Neilson.* Edited by A. R. Chisholm. Sydney: Angus & Robertson, 1965.

Nickerson, J.S. *Homage to Malthus.* Port Washington, New York: Kennikat Press, 1975.

Northbourne, Lord. *Look to the Land.* London: Dent, 1940.

Oldeman, L.R. "The Global Extent of Soil Degradation." *Soil Resilience and Sustainable Land Use: Proceedings of a Symposium Held in Budapest, 28 Sept–2 Oct 1992.* Edited by D. J. Greenland and I. Szabolcs. Wallingford, UK: CAB International, 1994: 561.

Ontario Agricultural Commission. *Report of the commissioners*, 3rd ed. City, Ontario Agricultural Committee, Year.

Paine J.A., et al, "Improving the nutritional value of Golden Rice through increased pro-vitamin A content," *Nature Biotechnology* 23 (2005): 482–7.

Parris, Leslie. *Landscape in Britain c. 1750–1850.* London: The Tate Gallery, 1973.

Patton, Andy and Shirley J.R. Madill. *Kim Adams*. Winnipeg: Friesen Printers, 1991.

Pearson, Craig. Email correspondence to B. Fisher and T. Podbury of the Australian Bureau of Agricultural and Resource Economics. Canberra, 12 January 2004.

Pearson, Craig. "Regenerative or semi-closed systems: a priority for twenty-first century agriculture," *BioScience* 57 (2007): 409–18.

Pearson, Craig, Geoff Cunningham, and David King. *A Plain English Guide to Agricultural Plants*. Melbourne: Longman Cheshire Pty Ltd, 1993.

Pearson, Craig J., and R.L. Ison. *Agronomy of Grassland Systems*, 2nd ed. Cambridge: Cambridge University Press, 1997.

Pearson, Craig J., H. Zuo, I. Valentine, and M. Unkovich. "Sustainability of Grazing Systems." *International Journal of Agricultural Sustainability* 1 (2003): 95–107.

Perry, T.M. *Australia's First Frontier: The Spread of Settlement in New South Wales, 1788–1829*. Melbourne: Melbourne University Press, 1963: 162.

Peterson, J., Fiona Cornwell and C.J. Pearson. *Chain Stocktake of Some Australian Agricultural and Fishing Industries* (Canberra: Bureau of Rural Sciences, 2000), 15.

Petzold, Andreas. *Romanesque Art*. New York: Harry N. Abrams, 1995.

Pretty, J.N., C. Brett, D. Gee, R.E. Hine, C.F. Mason, J.I.L. Morison, et al. "An assessment of the total external costs of UK agriculture," *Agricultural Systems* 65, no. 2 (2000): 113–36.

Proulx, Annie. *That Old Ace in the Hole*. New York: Scribner, 2002.

Quinton, Sarah, ed. *Stitching Women's Lives: Sujuni and Khatwa from Bihar, India*. Toronto: Textile Museum of Canada, 2000.

Ranney, D.C. "Mobile capital and economic development planning." *Journal of Planning Education & Research* 20 (2001): 281–92.

Rapport, D.J., W.S. Fyfe, R. Costanza, J. Spiegel, A. Yassi, G.M. Bohm, et al. "Ecosystem health: definitions, assessment, and case studies." In *Our Fragile World: Challenges and Opportunities for Sustainable Development*, ed. M.K. Tolba (Oxford: EOLSS Publishers, 2001): 21–42.

Rapport, D.J., and W.G. Whitford. "How ecosystems respond to stress." *BioScience* 49 (1999): 193–203.

Reid, Dennis. *Our Own Country Canada: Being an Account of the National Aspirations of the Principal Landscape Artists in Montreal and Toronto: 1860–1890*. Ottawa: The National Gallery of Canada, 1979.

Reitzenstein, Reinhard. *escarpment, valley, desert* [exhibition catalogue]. Hamilton: Art Gallery of Hamilton, 2002.

Robinson, Jancis, ed. *The Oxford Companion to Wine*. New York: Oxford University Press, 1994.

Rockefeller Foundation. *Memorandum of understanding between Government of Mexico and Rockefeller Foundation,* 14 Sept 1965. Sleepy Hollow, New York: Rockefeller Archive Center.

Rockefeller Foundation. *Partnership agreement between the government of Mexico and Rockefeller Foundation,* 12 April 1966. Sleepy Hollow, New York: Rockefeller Archive Center.

Rockefeller Foundation. *President's Ten-year Review and Annual Report 1971*. Sleepy Hollow, New York: Rockefeller Foundation, 1971.

Röling, N. "Platforms for Decision-making about ecosystems." In *The Future of the Land,* edited by J.O. Fresco et al. New York: John Wiley & Sons, 1994: 385–93.

Rosener, W. *Peasants in the Middle Ages*. Urbana: University of Illinois Press, 1992.

Ross, Alexander M. *The College on the Hill: A History of the Ontario Agricultural College, 1874–1974*. Toronto: Copp Clark Publishing, 1974.

Ryan, Terrence. *Cape Dorset Prints/Estampes*. Canada: C.F. Haughton Ltd., 1973.

Schama, Simon. *Landscape and Memory*. Toronto: Random House, 1995.

Schoolcraft, Henry R. *Historical and Statistical Information, Respecting the History, Condition and Prospects of the Indians of the United States*. Philadelphia, 1851–57.

Schweitzer, A. *Peace or Atomic War?* New York: Henry Holt and Company, 1958.

Skole, D. and C. Tucker. "Tropical deforestation and habitat fragmentation in the Amazon." *Science* 260 no. 5116 (1993): 1905–10.

Slaughter, T.P. *Exploring Lewis and Clark: Reflections on Men and Wilderness*. New York: Random House, 2003.

Slive, Seymour. *Dutch Painting 1600–1800*. Connecticut: Yale University Press, 1995.

Soil Association. "A Brief History of the Soil Association." Soil Association website, www.soilassociation.org/web/sa/saweb.nsf/Library?OpenForm&Cat=_History.

Somerville, M.A., and D. Rapport, editors. *Transdisciplinarity: Recreating Integrated Knowledge*. Montreal: McGill-Queen's University Press, 2000.

Srinivasan, Viji. *Newer Horizons*. India: Kolor Flow Madras, 1999.

Statistics Canada. *Area of land in farm holdings, census data, Canada and by province, 1871 to 1971*, www.statcan.ca/english/freepub/11-516-XIE/sectionm/sectionm.htm#Gen%20Stat (accessed 2003)

Stern, V.M., R.F. Smith, R. van den Bosch, and K.S. Hagen. "The integrated control concept." *Hilgardia* (1959): 81–101.

Stiglitz, Joseph E. *Globalization and Its Discontents*. New York: W.W. Norton & Company, 2002.

Strong, Roy. *The Spirit of Britain: A Narrative History of the Arts*. London: Pimlico, 2000.

Swiatowska, Bogna, ed. *Between Us*. Warsaw: Wydawnictwo Promocyjne Miedzy Nami, vol. 5, 2003.

Taylor, Patrick. *Period Gardens: New Life for Historic Landscapes*. New York: Atlantic Monthly Press, 1991.

Terry, Martin. *Masterpieces of Australian Painting*. Sydney: Bay Books Pty. Ltd., 1986.

Thackway, R. and I.D. Cresswell. "A bioregional framework for planning the national system of protected areas in Australia." *Natural Areas Journal* (1997): 241–7.

Themeda ed. *Australian Dryland Salinity Assessment 2000: Extent, impacts, processes, monitoring and management options*. Braddon, ACT, Commonwealth of Australia: National Land and Water Resources Audit, January 2001, http://audit.ea.gov.au/ANRA/land/docs/national/Salinity_Contents.html.

Thibier, M. Stabilization of Numbers of in vivo Collected Embryos in Cattle but Significant Increases of in Vitro Bovine Produced Embryos in Some Parts of the World. Data Committee Annual Report. *IETS Newsletter* 22 (2004): 12–19.

Thomson, Grace. *Eleanor Bond: Work Station*. Winnipeg: Gallery I.I.I., University of Manitoba, 1989.

Thorp, Nigel. *The Glory of the Page: Medieval & Renaissance Illuminated Manuscripts from Glasgow University Library.* Glasgow: Glasgow University Library, 1987.

Todorovová, J., ed. *Národní Muzeum Praha: A guidebook: The Náprestek Museum of Asian, African and American Cultures.* Prague: Národní Muzeum, 2000.

Toronto Alternative Art Fair International. *Canada's Only Alternative Art Fair in the Heart of Toronto's West Queen West Art & Design District.* Toronto: TAAFI, 2005.

Turner, F.J. *Rise of the New West.* New York: Harper & bros., *circa* 1906.

Turner, F. J. *Significance of the Frontier in American History.* Ann Arbor: University Microfilms, 1966 [1928].

United Nations General Assembly, Rio Declaration on Environment and Development. *Report on the United Nations Conference on Environment and Development* (Rio de Janeiro: United Nations, 3–14 June 1992), www.unep.org/Documents.Multilingual/Default.asp?DocumentID=78&ArticleID=1163.

United States Department of Agriculture. *National Agricultural Statistics Service-Agricultural Statistics Database,* www.nass.usda.gov.

Van Zanden, J.L. "The first green revolution. The growth of production and productivity in European agriculture." *Economic History Review* 44 (1991): 215–39.

Venetoulis, J., Dahlia Chazan, and Christopher Gaudet. "Ecological footprint of nations." In *Redefining Progress.* Oakland, CA: Redefining Progress, 2004.

Wackernagel, Mathis, and William E. Rees. *Our Ecological Footprint: Reducing Human Impact on the Earth.* Gabriola Island, BC: New Society Publishers, 1996.

Wackernagel, M., C. Monfreda, and D. Deumling. "Ecological footprint of nations." *Redefining Progress.* Oakland, California: Redefining Progress, 2004. www.rprogress.org.

Walther, Ingo F., editor. *Masterpieces of Western Art: A history of art in 900 individual studies from the Gothic to the Present Day.* Köln: Taschen, 2002.

Waltner-Toews, David. "A Bill from the Power Company." In *The Fat Lady Struck Dumb*. London, ON: Brick Books, 2000.

Weaver, J.C. "Beyond the Fatal Shore: Pastoral Squatting and the Occupation of Australia, 1826 to 1852." *The American Historical Review*, 101 (4), 1996: 981–1007.

Weida, William J. Comments on the permit materials submitted by DGH and the Taiwan Sugar Corporation concerning economic development and a farrow-to-finish hog operation in the county of Flagstaff, Alberta, Canada, www.factoryfarm.org/docs/hardestyappealmaster1000.doc.

Wilderness Society. "Who we are." Wilderness Society website, 2006, www.wilderness.org/AboutUs/who.cfm.

Wilkes, A.R. "Arable Farming after the Napoleonic wars." *Agricultural History Review* 28 (1980): 90–103.

Wilson, Rev. John M. *The Rural Cyclopedia* or *General Dictionary of Agriculture and of the Arts, Sciences, Instruments and Practices*. Edinburgh: Fullarton & Co., 1854.

Wing, S., D. Cole, and G. Grant. "Environmental injustice in North Carolina's hog industry." *Environmental Health Perspectives* 108 no. 3 (2000): 225–31.

Wolf, R. *Organic farming: Yesterday's and tomorrow's agriculture*. Emmaus, Pennsylvania: Rodale Press, 1977.

Woodwell, G.M. "The functional integrity of normally forested landscapes: A proposal for an index of environmental capital." *Proceedings of the National Academy of Sciences* 99 (2002): 13600–5.

World Economic Forum. *2001 Environmental Sustainability Index: An Initiative of the Global Leaders of Tomorrow Environment Task Force*. Davos, Switzerland, 2001. http://sedac.ciesin.org/es/esi/ESI_01a.pdf.

World Resource Institute. "WRI findings underscore growing crisis of confidence in international institutions" [news release]. Washington, DC: World Resource Institute, 5 June 2003, http://governance.wri.org/newsrelease_text.cfm?NewsReleaseID=247.

Wright, Ronald. *A Short History of Progress*. Toronto: House of Anansi Press Inc., 2004.

Yarlow, Loretta, et al., ed. *Rodney Graham: Works from 1976 to 1994*. Toronto: Art Gallery of York University, 1994.

Young Man, Alfred. *North American Indian Art: It's a Question of Integrity*. Kamloops: Kamloops Art Gallery, 1998.

Zadoks, J.C. *Development of Farming Systems: Evaluation of the Five Year Period 1980–1984*. Netherlands: Centrum voor Landbouwpublikaties en Landbouwdocumentatie, 1989.

Zavitz, E.J. *Fifty Years of Reforestation in Ontario*. Toronto: Ontario Department of Lands and Forests, 1961.

INFORMATION ON ILLUSTRATIONS

Figure 1 The Nymphaeum at Villa D'Este, Cernobbio, Italy
Photo: Villa d'Este, Cernobbio, Italy <www.villadeste.it>

Figure 2 *Cerne Abbas Giant*, Dorset, England, circa AD 180–193 (disputed)
Photo: Martin Gray <www.sacredsites.com>

Figure 3 Master of the Middle Rhineland
Virgin in a Garden, circa 1410
(tempera on wood)
Collection: Städel Museum, Frankfurt am Main
Photo: © Blaudel/Gnamm – ARTOTHEK <www.artothek.de>

Figure 4 Pieter Brueghel the Elder
Der Frühling (*Spring*), 1565
(brown ink on paper)
Collection: Albertina, Vienna <www.albertina.at>

Figure 5 Antoni Waterloo after Johannes Ruisher
Village on the River, circa 1640
(drypoint and etching)

Collection: Gift of Andrew and Helen Brink, in memory of R. Alexander Brink and Edith Margaret Whitelaw Brink, 2007, University of Guelph Collection at the Macdonald Stewart Art Centre <www.msac.ca>

Figure 6 *October*, from the Fécamp Psalter, circa 1180
(illuminated manuscript 76 F 13, folio 10 verso)
Collection: Koninklijke Bibliotheek (National Library of the Netherlands)
<www.kb.nl>

Figure 7 Pieter Brueghel the Elder
Summer – Peasant Harvest, 1568
(pen and brown ink on paper)
Collection: Hamburger Kunsthalle, Hamburg, Germany
Photo: Elke Walford; Bildarchiv Preussischer Kulturbesitz / Art Resource, NY
<www.hamburger-kunstalle.de> <www.bpk-images.de>

Figure 8 Artist unknown
A *View of Colebrooke Dale in Shropshire and the Adjacent Country*, 1776
(hand-coloured copper engraving)
Collection: Art Centre Purchase, 2004, Macdonald Stewart Art Centre Collection <www.msac.ca>

Figure 9 Leonard Knyff (illustrator), Johannes Kip (engraver)
Badminton in the County of Gloucester, one of the Seats of the Most Noble & Potent Prince Henry, Duke of Beaufort, 1724
(engraving)
Collection: © British Library Board. All Rights Reserved. (191.g.13)
<www.bl.uk>
Photo: *Nouveau Théâtre de la Grand Bretagne*, Vol. I, Plate 9

Figure 10 Paulus Potter
Cows Reflected in the Water, 1648
(oil on panel)
Collection: Royal Cabinet of Paintings, Mauritshuis, The Hague
<www.mauritshuis.nl>

Figure 11 Allart van Everdingen
The Hayrick with the Movable Roof, 17th century
(etching)
Collection: Gift of Andrew and Helen Brink, in memory of R. Alexander Brink and Edith Margaret Whitelaw Brink, 2007, University of Guelph Collection at the Macdonald Stewart Art Centre <www.msac.ca>

Figure 12 T. Hearne (illustrator), Benjamin Thomas Pouncy (engraver)
No. 1 and *No. 2*, 1794
(etching)
Collection: © British Library Board. All rights reserved. (1043.l.29) <www.bl.uk>
From: Richard Payne Knight, *The Landscape, a Didactic Poem in Three Books* (London: W. Bulmer and Co., 1794)

Figure 13 William Kurelek
Plowing, not dated
(acrylic, graphite, coloured pencil on masonite)
Collection: Doris Priddle

Figure 14 Samuel de Champlain
A Plan for the proposed settlement at Isle Saint-Croix on the south shore of the Bay of Fundy, 1613
(ink sketch)

Figure 15 Capt. S. Eastman (illustrator), James Smillie (engraver)
Gathering Wild Rice, 1851–1857
(engraving)
Collection: Toronto Public Library <www.torontopubliclibrary.ca>
From: Henry Rowe Schoolcraft, *Historical and Statistical Information, Respecting the History, Condition and Prospects of the Indian Tribes of the United States*. Philadelphia, Volume 3, Plate 4 (1851–57)

Figure 16 *Fish Pond at Monticello*
Collection: Monticello / Thomas Jefferson Foundation, Inc. <www.monticello.org>
Photo: Robert C. Lautman

Figure 17 *Vegetable Garden at Monticello*
Collection: Monticello / Thomas Jefferson Foundation, Inc.
<www.monticello.org>
Photo: Robert C. Lautman

Figure 18 André Thevet
Strange Beast, 1557
(woodcut)
Collection: Toronto Public Library <www.torontopubliclibrary.ca>
From: *Les singularitéz de la France Antarctique*, Paris: Chez les heritiers de Maurice de la Porte, 1557 (1st edition), Repr. opp. pg. 145

Figure 19 Peter Rindisbacher
Assiniboine Hunting on Horseback, 1833
(watercolour with glazing over graphite underdrawing on paper)
Collection: Amon Carter Museum, Fort Worth, Texas (1966.50)
<www.cartermuseum.org>

Figure 20 Paul Kane
A Buffalo Pound, 1846
(oil on canvas)
Collection: Royal Ontario Museum (ROM 912.1.33) <www.rom.on.ca>
Photo: With permission of the Royal Ontario Museum © ROM

Figure 21 Ebenezer Birrell
Good Friends, circa 1830
(oil on canvas)
Collection: Art Gallery of Hamilton, Gift of Mrs. R.N. Steiner in memory of her mother, Mrs. L.C. Dillon, 1965 <www.artgalleryofhamilton.com>

Figure 22 Oscar Bluemner
Space Motive, a New Jersey Valley, circa 1917–18
(oil on canvas)
Collection: Whitney Museum of American Art, New York; Purchase, with funds from Mrs. Muriel D. Palitz 78.2 <www.whitney.org>
Photo: Sheldon C. Collins

Figure 23 Alfred Russell Colman
A.R. Colman Veterinary Surgeon, 1876
(gouache and watercolour)
Collection: Gift of Mr. and Mrs. W.B. Fox, 1978, University of Guelph Collection at the Macdonald Stewart Art Centre <www.msac.ca>

Figure 24 Charles Manly
The Ontario Agricultural College, Guelph, Canada, 1906
(watercolour)
Collection: Ontario Agricultural College Purchase, 1906, University of Guelph Collection at the Macdonald Stewart Art Centre <www.msac.ca>

Figure 25 *Government Agricultural Establishment, Castle Hill*, circa 1806
(watercolour)
Collection: Mitchell Library, State Library of New South Wales <www.sl.nsw.gov.au>

Figure 26 Joseph Lycett
Raby, a farm belonging to Alexander Riley Esq., New South Wales, 1825
(etching and aquatint, printed in black ink from one copper plate, hand-coloured)
Collection: National Gallery of Australia, Canberra <www.nga.gov.au>

Figure 27 After Andrew Putnam Hill
George Hoag's Steam Threshing Outfit and Crew Setting a New One-day World's Record, 1878
(lithograph)
Collection: F. Hal Higgins Library of Agricultural Technology, University of California, Davis <www.lib.ucdavis.edu> / The Smithonian Institution

Figure 28 Russell Drysdale
A Man Feeding his Dogs, 1941
(oil on canvas)
Collection: Queensland Art Gallery, Gift of C.F. Viner-Hall, 1961 <www.qag.qld.gov.au>

Figure 29 John Glover
A view of the artist's house and garden, in Mills Plain, Van Diemen's Land, Deddington, Tasmania, 1835
(oil on canvas)
Collection: Art Gallery of South Australia, Morgan Thomas Bequest Fund, 1951 <www.artgallery.sa.gov.au>

Figure 30 Thomas Hart Benton
Departure of the Joads, 1939
(lithograph)
Collection: Ball State University Museum of Art, Gift of Ned and Gloria Griner, 2005 (2005.43.4)
<www.bsu.edu/artmuseum/>

Figure 31 Margaret Bourke-White
World's largest dam across the Dnieper River under construction, 1931
(silver gelatin print)
Collection: Time & Life Pictures
Photo: Margaret Bourke-White / Stringer
Credit: Time & Life Pictures / Getty Images
<http://editorial.gettyimages.com>
© Time & Life Pictures

Figure 32 Jean-François Millet
The Gleaners, 1857
(oil on canvas)
Collection: Musée d'Orsay, Paris, France <www.musee-orsay.fr>
Photo: Jean Schormans; Réunion des Musées Nationaux / Art Resource, NY

Figure 33 Käthe Kollwitz
Pflugzieher und Weib (Klipstein 61) [*Plowmen and Woman*], 1902
(lithograph)
Collection: Gift of Fine Art Printmaking Students, 1974, University of Guelph Collection at the Macdonald Stewart Art Centre <www.msac.ca>
Photo: © Estate of Käthe Kollwitz / SODRAC (2007)

Figure 34 Théodore Géricault
Pity the Sorrow of a Poor Old Man, 1821
(lithograph)
Collection: The Baltimore Museum of Art, Purchase with exchange funds from Gift of Dr. Noah Dorsky, Gift of Barry Leon, Gift of Dr. M.H. Mandlebaum, and Gift of Anthony Phillips BMA 1978.121
<www.artbma.org>

Figure 35 Honoré Daumier
Les savants dans l'exercise de leurs fonctions [*Skilled gentlemen in an exercise of their functions*], 1853
(lithograph)
Collection: Robert D. Farber University Archives and Special Collections Department, Brandeis University Libraries & Trustman Daumier Collection
<www.dembitz.mainlib.brandeis.edu>

Figure 36 Thomas Weaver
A Short Horned Heifer, Seven Years Old, circa 1811–19
(engraving)
Collection: Gift of Stewart and Letty Bennett, donated by the Ontario Heritage Foundation, 1988, University of Guelph Collection at the Macdonald Stewart Art Centre <www.msac.ca>

Figure 37 (Top) Artist unknown
Avenue of Champions, 1907
(opaque watercolour)
Collection: The Fair Publishing House, Inc. Collection, Michigan State University Museum <www.msu.edu>

Figure 37 (Bottom) W. Layman
Woman with Fruit (Fair Scene), circa 1910
(opaque watercolour)
Collection: The Fair Publishing House, Inc. Collection, Michigan State University Museum <www.msu.edu>

Figure 38 David Nasby
Booth at Fall Fair, Wingham, Ontario, Canada, 1971
(silver gelatin print)
Collection of the artist

Figure 39 (Top) *The Barn of the Provident Master*
From: J. Russell Manning, *The Illustrated Stock Doctor and Livestock Encyclopedia*, pp. 149–50

Figure 39 (Centre) *Farmer Unthrift's Barn*
From: J. Russell Manning, *The Illustrated Stock Doctor and Livestock Encyclopedia*, pp. 149–50

Figure 39 (Bottom) *Farmer Unthrift's Home*
From: J. Russell Manning, *The Illustrated Stock Doctor and Livestock Encyclopedia*, pp. 149–50

Figure 40 Charles Demuth
My Egypt, 1927
(oil and graphite on composition board)
Collection: Whitney Museum of American Art, New York; Purchase, with funds from Gertrude Vanderbilt Whitney 31.172 <www.whitney.org>
Photo: Sheldon C. Collins

Figure 41 A.J. Casson
"V" is for Victory, circa 1939–45
(oil on masonite)
Collection: Gift of the Lucas Verity Corporation, Massey-Ferguson Archives, 1999, University of Guelph Collection at the Macdonald Stewart Art Centre <www.msac.ca>

Figure 42 Victor Ivanov and Olga Burova
A Tractor in the Field is Worth a Tank in Battle, 1942
(propaganda poster)
Collection: Russian Poster Collection, University of Guelph Library <www.lib.uoguelph.ca>

Figure 43 Alan Caswell Collier
Ploughing in Tibet, not dated
(oil on canvas board)
Collection: Gift of the Lucas Verity Corporation, Massey-Ferguson Archives, 1999, University of Guelph Collection at the Macdonald Stewart Art Centre <www.msac.ca>

Figure 44 Anthony Kingscote
Ploughing the Rice Paddy, circa 1968
(ink and wash on paper)
Collection: Gift of the Artist, 1977, University of Guelph Collection at the Macdonald Stewart Art Centre <www.msac.ca>

Figure 45 Otis Dozier
Grasshopper and Farmer, 1937
(lithograph)
Collection: Dallas Museum of Art, gift of the artist <www.dallasmuseumofart.org>

Figure 46 Reinhard Reitzenstein
No Title, 1987
(trees inverted and peeled)
Collection: Top Sail Island, Sault Ste. Marie, 1987–88
Photo: Reinhard Reitzenstein

Figure 47 Robert Smithson
Glue Pour, Vancouver, B.C., 1969
(aluminum drum, glue)
Collection: Estate of Robert Smithson, © VAGA (New York) / SODART (Montreal) 2006
Photo: Courtesy James Cohan Gallery, New York

Figure 48 Andy Warhol
Tuna Fish Disaster, 1963
(silkscreen ink and silver paint on linen)
Collection: © The Andy Warhol Foundation for Visual Arts / CORBIS / SODRAC (2007)
Photo: <www.corbis.com> <www.warholfoundation.org>

Figure 49 Robert Howson
Boo Moo, 2002
(digital print)
Collection: Purchased with funds raised by the Art Centre Volunteers and with the financial support of the Canada Council for the Arts Acquisition Assistance Program, 2003, Macdonald Stewart Art Centre Collection <www.msac.ca>

Figure 50 Police warning to keep out at Burnside Farm, Northumberland, where the UK foot-and-mouth crisis is thought to have originated.
Photo: 15 March 2001, Associated Press <www.ap.org>

Figure 51 Disposal of cattle at Lockerbie, Scotland.
Photo: 3 March 2001, Associated Press <www.ap.org>

Figure 52 Diane Maclean
Wether, 2001
(Clashach sandstone and bronze)
Collections: 1) Timespan, Helmsdale, Scotland <www.timespan.org.uk>
2) John Manser, Chisenbury Priory, Wiltshire
Photo: Diane Maclean

Figure 53 Margaret Bourke-White
Rendering In Chicago, 1930
(silver gelatin print)
Collection: Time & Life Pictures
Photo: Margaret Bourke-White / Stringer
Credit: Time & Life Pictures / Getty Images
<http://editorial.gettyimages.com>
© Time & Life Pictures

Figure 54 David Nasby
Max Storey, Auctioneer, 1972
(silver gelatin print)
Collection of the Artist

Figure 55 Rodney Graham
Napoléon Linden, Marbais, 1992–93
(C-print)
Collection: Group Lhoist (Belgium)
Photo: Courtesy Scott Livingstone, Rodney Graham Studio

Figure 56 *Harrowsmith*, No. 20, July 1979
Collection: Harrowsmith Country Life R/T of Harrowsmith Magazine
<www.harrowsmithcountrylife.ca>

Figure 57 Kim Adams
Earth Wagons, 1989–91
(HO model parts, utility trailers, plastic landscape material, electrical components)
Collection: The Winnipeg Art Gallery, Acquired with funds from The Winnipeg Art Gallery Foundation Inc. and the Volunteer Committee to The Winnipeg Art Gallery, Accession # G-91-293 <www.wag.mb.ca>
Photo: Ernest Mayer, The Winnipeg Art Gallery; © Kim Adams

Figure 58 Lee Friedlander
Idaho, 1972
(silver gelatin print)
Collection: Lee Friedlander Photographs, pl. 83, Friedlander, MoMA, pl. 147
Photo: © Lee Friedlander, courtesy Fraenkel Gallery, San Francisco
<www.fraenkelgallery.com>

Figure 59 Eleanor Bond
Activity in the Inner Harbour is Regenerated by the World Botanical Garden, Constructed with Recycled Materials from the Glass City, 1995
(oil on canvas, unstretched)
Collection: National Gallery of Canada, Ottawa, Gift of the Artist, Montreal, 2002 <www.gallery.ca>
Photo: © National Gallery of Canada / CARCC, 2007

Figure 60 Edward Burtynsky
Nickel Tailings #34, Sudbury, Ontario, 1997
(chromogenic colour print)
Photo: Courtesy of the artist and Nicholas Metivier Gallery, Toronto

Figure 61 Orest Semchishen
Glen and Hilda Cole, Near Coronation, Alberta, 1980
(silver gelatin print)
Collection of the artist

Figure 62 Arthur Boyd
Irrigation Lake, Wimmera, 1950
(resin and tempera on composition board)
Collection: National Gallery of Victoria, Melbourne, Australia,
Purchased, 1950 <www.ngv.vic.gov.au>

Figure 63 *Cow Dung*
(sujuni textile, 200 stitches per square inch)
Drawing by Nirmila H.
Embroidered appliqué by women in Bihar, India
Collection: Skye Morrison
Photo: Courtesy Debbie Adams

Figure 64 Edward Lear
The Old Man of the East (from *A Book of Nonsense*), circa 1866
(silkscreen in colours)
Collection: The Metropolitan Museum of Art, Gift of Mr. and Mrs.
Bryan Holme, 1984 (1984.1056.41) <www.metmuseum.org>
Photo: © The Metropolitan Museum of Art

Figure 65 Claes Oldenburg
Floor Burger, 1962
(canvas filled with foam rubber and cardboard boxes, acrylic paint)
Collection: Art Gallery of Ontario, Toronto, Purchase 1967
<www.ago.org>
Photo: © Claes Oldenburg and Coosje van Bruggen, New York

Figure 66 John Greer
Three Grains of Wheat, 1991
(bronze)
Collection: Commissioned with funds donated by Isabel McLaughlin in memory of Norah McCullough, 1993, Macdonald Stewart Art Centre Collection <www.msac.ca>

Figure 67 Edward Burtynsky
Chicken Packaging Plant #1, Toronto, 1983
(chromogenic colour print)
Photo: Courtesy of the Artist and Nicholas Metivier Gallery, Toronto

Figure 68 David Hlynsky
Window, 3 Loaves of Bread, Poland, 1988
(colour photograph)
Collection of the artist
Photo: David Hlynsky

Figure 69 Thomas Grünfeld
misfit (cow/ostrich), 1997
(taxidermy)
Collection: Private collection, USA
Photo: © Thomas Grünfeld / SODRAC (2007)

Figure 70 *Science: American Association for the Advancement of Science*, Vol. 211, No. 4479 (23 January 1981)
Photo: J. Messineo, Fort Collins, Colorado <www.sciencemag.org>

Figure 71 Fred Williams
Saplings, Mittagong, 1961–62
(gouache)
Collection: National Gallery of Australia, Purchased from Gallery admission charges, 1983, Accn. No. NGA 83.2989.22
<www.nga.gov.au>
© Estate of Fred Williams

Figure 72 Orest Semchishen
Springwater, Saskatchewan, 1983
(colour photograph)
Collection of the artist

Figure 73 Tomek Sikora
Georgetown Autos, 2003
(transfer on card)
Collection of the artist <www.tomeksikora.com>
Photo: Tomek Sikora; Marzena Inc.

Figure 74 FASTWÜRMS
ex ovo omnia, 2000
(mixed media)
Collection: Commissioned through the Florence G. Partridge Fund in consultation with the Ontario Agricultural College, and with financial support from the Canada Council for the Arts Acquisition Assistance Program, 2000, Macdonald Stewart Art Centre Collection <www.msac.ca>

Figure 75 Noel Harding
The Elevated Wetlands, 1997
(garden material, concrete)
Sponsored by Plastics & Art, installed by the Don River, Toronto
Collection of the artist <www.noelharding.ca>
Photos: Noel Harding

Figure 76 Gu Xiong
The Sickle and the Cell Phone, 2002
(bronze)
Collection: Commissioned with funds donated by Ann Oaks, with financial support from the Canada Council for the Arts Acquisition Assistance Program, 2002, Macdonald Stewart Art Centre Collection <www.msac.ca>

Figure 77 Ernesto Apomayta
A Hard Day's Toil, 2004
(natural inks on cotton paper)
Collection of the artist <www.apomaytaart.com>
Photo: Ernesto Apomayta

Figure 78 Rodney Graham
Weather Vane, 2002
(black enamelled stainless steel)
Collection: Purchased with funds donated by Helen Brimmell, 2002, Macdonald Stewart Art Centre Collection <www.msac.ca>

Figure 79 Kananginak Pootoogook
Skinned Caribou, 1973
(stonecut and stencil, 13/50)
Collection: Purchased with funds donated by Blount Canada Ltd., with assistance from The Canada Council, 1981, Macdonald Stewart Art Centre Collection <www.msac.ca>

Figure 80 Emily Kam Kngwarray
Emu Story, 1989
(synthetic polymer paint on canvas)
Collection: National Gallery of Victoria, Melbourne, Australia, Purchased from admission funds, 1990 <www.ngv.vic.gov.au>

Figure 81 Antoni Waterloo
The Entrance to a Forest with a Small Wooden Bridge, circa 1660
(etching)
Collection: Gift of Andrew and Helen Brink, in memory of R. Alexander Brink and Edith Margaret Whitelaw Brink, 2007, University of Guelph Collection at the Macdonald Stewart Art Centre <www.msac.ca>

Figure 82 David Hoffos
Steamed Egg Custard with Vegetables, 2005
(shadow box with laser print and magnifying lens)
Collection: Canadian Photographic Portfolio Society (*Japanese Cooking*, a suite of 8 shadow boxes, edition of 10) <www.cppsonline.com>
Photo: David Hoffos

Figure 83
Southern Ojibway, Artist unknown
Lac Seul Master Scroll with Bear emphasis and herbal references, not dated
(natural pigment on parchment)
Collection: Starratt Collection
Photo: University of Toronto Press <www.utpress.utoronto.ca>
From: Selwyn Dewdney, *The Sacred Scrolls of the Southern Ojibway*, Figure 155 (Published for the Glenbow-Alberta Institute by the University of Toronto Press, 1975)

Figure 84 Christo and Jeanne-Claude
The Gates, Central Park, New York City, 1979–2005
(steel, vinyl tube, cast aluminum, nylon thread)
Collection of the artists
Photo: Wolfgang Volz <www.wolfgangvolz.com>; © Christo 2005, Central Park, New York, USA, 02/2005 (Bildtechnik: Farbprofil sRGB); LAIF <www.laif.de>; TCS <www.thecoverstory.com>

Figure 85 Bardayal (Lofty) Nadjamerrek
Ngalyod, the Rainbow Serpent, 1998
(natural ochres on paper)
Private collection

Figure 86 Lily Karadada
Wandjina, 1990
(earth pigments and natural binder on canvas)
Collection: National Gallery of Victoria, Melbourne, Australia, Purchased through The Art Foundation of Victoria with the assistance of The Marjory and Alexander Lynch Endowment, Governors, 1990 (0.157–1990) <www.ngv.vic.gov.au>

Figure 87 Irene Avaalaaqiaq
Willow Bushes, 1970s
(graphite and coloured pencil on paper)
Collection: Purchased with funds donated by Blount Canada Ltd., 1993, Macdonald Stewart Art Centre Collection <www.msac.ca>

ABOUT THE AUTHORS

CRAIG PEARSON

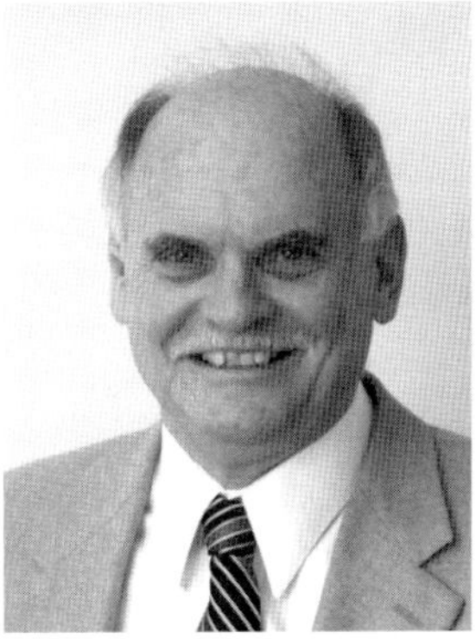

Craig Pearson is professor of Agricultural and Environmental Policy at the University of Guelph, Canada. He holds degrees from the University of Western Australia, the University of Guelph, and Macquarie University, and holds a Senior Management Certificate from Harvard. He was dean of the Ontario Agricultural College at Guelph from 2001 to 2007 and has served as chief scientist for Australia's Department of Agriculture, Fisheries and Forestry; as adjunct professor at the Australian National University; and as member of the board of the Cooperative Research Centre for Greenhouse Accounting in Canberra, Australia. He has also held the positions of executive dean, Faculty of Natural Resources, Agriculture and Veterinary Science; pro-vice-chancellor at the University of Queensland; professor of Agronomy, University of Sydney; visiting professor in Environmental and Information Sciences, Charles Sturt University; and coordinator of Tropical Crops Research at the University of Western Australia. Pearson is an elected fellow of the Australian Institute of Agricultural Science and Technology. He has authored ten books and more than one hundred papers. His primary research interest is agricultural and environmental policy and sustainability, although his career has spanned plant physiology, farming systems and participative technology transfer.

JUDITH NASBY

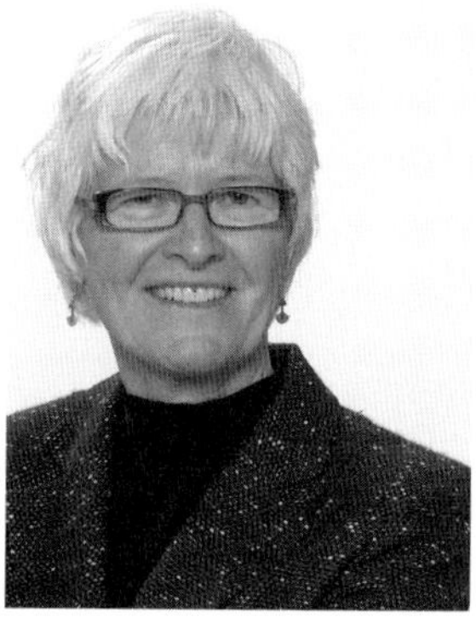

Judith Nasby has been a curator and public art gallery director for more than thirty years. As curator at the University of Guelph, she directed the University Art Gallery until 1975. As director of the Macdonald Stewart Art Centre in Guelph, she oversees one of the most comprehensive sculpture parks in Canada and a permanent art collection of over six thousand works. She has curated over one hundred exhibitions and has written fifty publications, including *Irene Avaalaaqiaq: Myth and Reality* (McGill-Queen's University Press, 2002) and *Rolph Scarlett: Painter, Designer, Jeweller* (McGill-Queen's University Press, 2004). She has lectured and toured exhibitions to Austria, Brazil, China, Czech Republic, Denmark, Hungary, Iceland, India, Japan, Panama, Poland, Venezuela, and the United States. In 2003 she was a distinguished international visiting scholar at the Center for the Study of American Material Culture at the University of Delaware. She is also an adjunct professor in the School of Fine Art and Music at the University of Guelph.

INDEX